GRANITE LANDFORMS

GRANITE LANDFORMS

C.R. TWIDALE

University of Adelaide, Adelaide, South Australia, Australia

ELSEVIER SCIENTIFIC PUBLISHING COMPANY
Amsterdam — New York — Oxford   1982

ELSEVIER SCIENTIFIC PUBLISHING COMPANY
Molenwerf 1
P.O. Box 211, 1000 AE Amsterdam, The Netherlands

Distributors for the United States and Canada:

ELSEVIER SCIENCE PUBLISHING COMPANY, INC.
52, Vanderbilt Avenue
New York, N.Y. 10017

Library of Congress Cataloging in Publication Data

Twidale, C. R.
    Granite landforms.

    Bibliography: p.
    Includes index.
    1. Granite. 2. Landforms. I. Title.
QE462.G7T9 1982   552'.3   82-13939
ISBN 0-444-42116-5

Printed in The Netherlands

Dedicated with admiration, gratitude
and affection to Edwin Sherbon Hills,
FRS, Emeritus Professor of Geology in
the University of Melbourne, and for
more than 30 years mentor, colleague,
critic and friend.

CONTENTS

List of figures   xi
Preface   xxi

PART I   INTRODUCTION

Chapter 1. Landforms developed on granitic rocks   1

Chapter 2. Characteristics of granitic rocks   26
          A.   Occurrences of granite   26
          B.   Definition and composition   36
          C.   Some physical characteristics   39
               (i)   General   39
               (ii) Fractures   43
                     (a) Orthogonal sets   43
                     (b) Rift, grain and hardway   46
                     (c) 'Bedded' granite   48
                     (d) Sheeting joints   49

Chapter 3. Weathering, with particular reference to granitic rocks   58
          A.   Physical weathering   60
          B.   Chemical attack   64
          C.   Initial breakdown   68
          D.   Factors influencing the weathering of granite   71
               (i)   Climate   71
               (ii) Rock composition   73
               (iii) Texture   79
               (iv) Partings   82

PART II   MAJOR FORMS AND ASSEMBLAGES

Chapter 4. Boulders   89
          A.   Subsurface exploitation of orthogonal fracture sets   89
          B.   Historical perspective   96
          C.   Types of peripheral weathering   111
          D.   Evacuation of debris   119
          E.   Disintegration of sheet structure   119

Chapter 5. Inselbergs   124
            A. Bornhardt characteristics   124
            B. Reasons for positive relief   125
            C. Subsurface initiation   135
            D. Environments of development   136
            E. Scarp retreat   138
            F. Evidence and argument   139
                  (i) Fracture controlled margins   139
                  (ii) Regional settings   139
                  (iii) Upland settings   139
                  (iv) Deep weathering, and contrast between hill and plain   141
                  (v) Incipient domes   142
                  (vi) Subsurface initiation of minor forms   143
                  (vii) Flared slopes and stepped inselbergs   144
                  (viii) Age of inselbergs   146
                  (ix) Occurrence in multicyclic landscapes   147
            G. Domical form   149
                  (i) Weathering   149
                  (ii) Sheet structure   150
                      (a) Exogenetic theories   150
                      (b) Endogenetic theories   154
            H. Other inselbergs   158
                  (i) Nubbins   161
                  (ii) Castle koppies   167

Chapter 6. All-slopes topography   177
            A. Distribution   177
            B. Origins   179

Chapter 7. Granite plains   186
            A. Buried and exhumed plains   187
            B. Etch plains   187
            C. Plains of subaerial (epigene) origin   189
                  (i) Pediments   190
                      (a) Mantled   190
                      (b) Rock pediments or platforms   193
                      (c) Origins   199
                  (ii) Peneplains   201
                  (iii) Relationship between peneplain and pediment   202
                  (iv) Pediplain and ultiplain   204
                  (v) Multicyclic and stepped assemblages   207

PART III   MINOR LANDFORMS

Chapter 8. Forms of gentle slopes   213

            A. Rock basins   214

                (i) Description   214

                (ii) Nomenclature   214

                (iii) Origin   217

                    (a) General comments   217

                    (b) Rate of development   219

                    (c) Differentiation of major types   221

                    (d) Initiation at the weathering front   227

            B. Pedestals   228

                (i) Description   228

                (ii) Origin   228

            C. Rock doughnuts   233

                (i) Description   233

                (ii) Origin   233

                    (a) Induration   233

                    (b) Water scour   234

                    (c) Relief inversion   234

                (iii) Evidence and argument   235

            D. Runnels or gutters   237

                (i) Description   237

                (ii) Origins   239

Chapter 9. Forms associated with steep slopes   243

            A. Flared slopes   243

                (i) Distribution   243

                (ii) Description and characteristics   244

                (iii) Origin   247

            B. Fretted basal slopes   257

            C. Rock platforms   258

                (i) Description and distribution   258

                (ii) Origin   260

            D. Scarp foot depressions   260

                (i) Description   260

                (ii) Origin   262

            E. The piedmont angle   265

            F. Grooves or flutings   267

                (i) Description   267

                (ii) Origins   270

                    (a) Structural factors   270

(b) Processes at work   270

(c) Coastal developments   274

(iii) Subsurface initiation   275

Chapter 10. Caves and tafoni   280

A. Caves associated with corestones and grus   280

B. Caves associated with fractures   281

C. Tafoni   281

(i) Description   281

(ii) Origin   288

(a) Initiation   288

(b) Development   292

(c) Visor   297

(iii) Case hardening and other coatings   297

Chapter 11. Split and cracked blocks and plates   301

A. Split rocks   301

B. Parted blocks   305

C. Polygonal cracking   306

(i) Description   307

(ii) Origin   311

(a) Previous work   311

(b) Analysis   313

(c) Possible explanations   315

D. Displaced slabs and blocks   317

(i) A-tents   317

(ii) Overlapping slabs   321

(iii) Displaced slabs   322

(iv) Wedges   325

(v) Origin of the forms   327

PART IV   OVERVIEW

Chapter 12. Discussion and Conclusions   330

References cited.   339

Index   359

LIST OF FIGURES

S.A. - South Australia; W.A. - Western Australia; N.T. - Northern Territory; N.S.W. - New South Wales; Q'd. - Queensland; R.S.A. - Republic of South Africa; U.S.A. - United States of America.

CHAPTER 1

1. Fault scarps, eastern side Sierra Nevada, California.
2 (a). Granite displaced by faulting, Berridale, N.S.W.
  (b). Granite displaced by faulting, eastern Dartmoor, England.
3. MacDonald Fault Scarp, N.W.T. Canada.
4. Half Dome, Yosemite, Sierra Nevada, California.
5. Pearson Islands, S.A.
6. Joints in sea cliffs, Land's End, England.
7. Shore platform of etch origin near Streaky Bay, S.A.
8 (a). Granite boulder, north Q'd.
  (b). Group of boulders, Palmer, S.A.
  (c). Granite boulder, Tampin, West Malaysia.
9. Map of Everard Range, S.A.
10. Bornhardts in Groothoekseberg, Namaqualand, R.S.A.
11. Groot Spitzkoppe, central Namibia.
12. Angolan inselberg landscape (Jessen).
13. Bornhardts near Nanutarra, W.A.
14. Bushman Surface with isolated bornhardt, central Namaqualand, R.S.A.
15. Granite domes, southern Algeria.
16. The Sugarloaf and other *morros*, Rio de Janeiro, Brazil.
17. Nubbin near Naraku, N.W. Q'd.
18. Castle koppie, eastern Zimbabwe.
19. Kilba Hills, northern Nigeria (Falconer).
20. All-slopes topography, northern Flinders Ranges, S.A.
21. Sheet structure, Pearson Islands, S.A.
22. Pan on Haytor, eastern Dartmoor, England.
23. Fluted and flared slope, Pildappa Rock, S.A.
24. Sheet tafoni, Ucontitchie Hill, S.A.
25. Bornhardt and fringing pediment, Ucontitchie Hill, S.A.
26. Rock pediment, Corrobinnie Hill, S.A.
27. Peneplain in granite, Corrigin, W.A.

28.   Flat granite plain, Meekatharra, W.A.

29.   Flared boulder on shore platform, Streaky Bay, S.A.

CHAPTER 2

1 (a). Structure of the earth.

  (b). Structure of upper crust.

2.   Distribution of shields, orogens and platforms.

3 (a). Laurentian Shield.

  (b). Baltic Shield.

4 (a). Batholiths in southeastern Australia.

  (b). Batholiths in Brittany, France.

5.   Lenticular plutonic masses, Sierra Nevada, California.

6.   Plan of and section through Vredefort dome, Transvaal.

7.   Linked plutons of southwestern England.

8.   Mt Hillers stock and Stewart Ridge laccolith, Utah.

9.   Contact of batholith and host sediments, British Columbia, Canada.

10.   Laccolith, Judith Mts, Montana.

11.   Phacolith, Shropshire.

12.   Section through Bushfeld Lopolith, Transvaal.

13.   Gneiss domes, southern Finland.

14.   Ring intrusion, southwestern Finland.

15.   Composition and classification of granitic rocks.

16.   Section through Sierra Nevada batholith, California.

17 (a). Thin section of medium-grained equigranular granite.

   (b). Porphyritic granite, South East of S.A.

   (c). Thin section of augen gneiss.

18.   Fracture patterns in batholith, according to H. Cloos.

19.   Orthogonal joints, Mt Bundey, N.T.

20.   Orthogonal joints in gabbro, northwest of S.A.

21.   Large-scale rhomboidal joint sets, central Labrador.

22.   Fracture patterns, French Guyana.

23.   Microscopic structure of rift and grain.

24.   Flaggy structure (a) Roughtor, Dartmoor, southwestern England.
        (b) Yosemite region, California.

25.   Orthogonal and flaggy joints, Heltor, Dartmoor, southwestern England.

26.   Sheet structure, Ucontitchie Hill, S.A.

27.   Sheet structure, Little Shuteye Pass, California.

28.   Section through Ucontitchie Hill, S.A.

29.   Steeply dipping sheeting joints on flared and fluted eastern slope
            Ucontitchie Hill, S.A.

30.   Dome near Tenaya Lake, California.

31.   Wedges at Ucontitchie Hill, S.A.   (a) *in situ*;   (b) displaced.

32.   Rock of Ages Quarry, Vermont.

CHAPTER 3

1.   Mesa with iron oxide-rich capping, Devil's Marbles, N.T.

2.   Plateaux capped by laterite, N.W. Q'd.

3.   Weathering fronts marked by iron oxide concentration in granitic rocks.

4.   Boulder with fire flaking, Albany, W.A.

5.   Pitted and fire flaked boulder, Tampin, West Malaysia.

6.   Clitter, Rocky Mts, Colorado.

7.   Frost riven blocks, Andorran Pyrenees.

8.   Secondary mineral development at crystal boundaries.

9.   Microfissures due to percussion.

10.   Flaking in granite, Palmer, S.A.

11.   Granite cliffs with talus, Andorran Pyrenees.

12.   Encounter Bay, S.A.

13 (a). Map and section, Pão de Assucar, southeastern Brazil.

   (b). Section through the Gavea, southeastern Brazil.

14.   Corcovado, southeastern Brazil.

15.   Mantelluccio, Corsica.

16.   Comparative resistance of granitic rocks in Pyrenees.

17.   Pitting at base of Pildappa Rock, S.A.

18.   Pitting at weathering front.

19.   Pitting at Mt Bundey, N.T.

20.   Exploitation of crystal boundary, Leeukop, Transvaal.

21.   Pegmatite cleft, Namibia.

22.   Haytor West, Dartmoor, southwestern England.

23.   Veins, Andorran Pyrenees.

24.   Quartz veins, northwestern W.A.

25.   Aplitic sill, Paarlberg, R.S.A.

26 (a). Pavement with rims bordering fractures, Namibia.

   (b). Detail of rims.

27.   Joint cleft, Kwaterski Rocks, S.A.

28.   Joint clefts, Paarlberg, R.S.A.

29.   Crazy paving, Devil's Marbles, N.T.

CHAPTER 4

1.   The Leviathan, northeastern Victoria.
2 (a). Corestones and grus near Lake Tahoe, California.
  (b). Corestones in grus, Snowy Mts, N.S.W.
  (c). Corestones and grus, Gemencheh, West Malaysia.
  (d). Corestones and grus, Mt Bundey, N.T.
  (e). Corestones and grus, Rocky Mts, Colorado.
3.   Two stage development of boulders.
4 (a). Weathered granite and vein, Rocky Mts, Colorado.
  (b). Weathered granite and veins, Sardinia.
5.   *Compayrés*, Palmer, S.A.
6.   Turrets in the Hoggar Mts, southern Algeria.
7 (a). Sketch of the Cheesewring, Cornwall (Jones).
  (b). Cheesewrings, Devil's Marbles, N.T.
8.   The Logging Stone, Cornwall (Jones).
9 (a). Perched blocks, Balancing Rocks, Zimbabwe.
  (b). Perched boulder, Dartmoor, England.
  (c). Perched boulder, central Texas.
  (d). Peyro Clabado, southern France.
10.   Cottage loaf, Devil's Marbles, N.T.
11 (a). Tombstones, eastern Mount Lofty Ranges, S.A.
   (b). Foliation slabs, Reynolds Range, N.T.
12.   Bowerman's Nose, eastern Dartmoor, England.
13.   Hassenfratz's site, southern Massif Central, France.
14.   De la Beche's diagram of corestone and grus.
15.   Logan's site at Palu Ubin, Singapore.
16.   Geikie's sketch of corestone in grus.
17.   Onion weathering, Snowy Mts, N.S.W.
18.   Intradosal zone near excavation.
19.   Curved joints on shore platform near Streaky Bay, S.A.
20.   Mineral banding, Snowy Mts, N.S.W.
21.   Inselbergs, Lake Chad region, central Africa.
22.   Development of barrel-shaped corestones.
23.   Diagram of weathering profile.
24.   Corestones and grus, Snowy Mts, N.S.W.
25.   Disintegrated sheet structure, Mauritania.
26.   Disintegrated sheet structure, Namibia.
27.   Disintegrated sheet, Little Wudinna Hill, S.A.
28.   Disintegrated sheet structure, Paarl, R.S.A.
29.   Disintegrated sheet structure, Little Rock, Enchanted Rock complex, Texas.-

CHAPTER 5

1.   Examples of plan shapes of inselbergs determined by fractures.

2.   Fracture-controlled valley with convex sidewalls, Zimbabwe.

3.   Bornhardts in multicyclic landscape, eastern Transvaal, R.S.A.

4.   Aerial view, Yosemite valley, California.

5.   Pic Parana, southeastern Brazil.

6.   Geology of Stone Mountain area, Georgia, U.S.A.

7.   Stock-bornhardt, central Namibia.

8.   Folded crystallines, Rio de Janeiro area.

9.   Sketch of Blackingstone Rock and quarry (Jones).

10.   Ucontitchie Reservoir, S.A.

11.   Distribution of stress in anticline.

12.   Compression in region affected by *en echelon* wrench faults.

13.   Distortion of cube during shearing.

14.   Two stage development of bornhardts.

15.   Bornhardts as remnants following scarp retreat.

16.   Major fractures, northwestern Eyre Peninsula, S.A.

17.   Alignment of residuals, north of Wudinna, S.A.

18.   Domes at various levels in landscape in central Namaqualand, R.S.A.

19.   Field sketch, Swakop valley, central Namibia.

20.   Artificially exposed dome, Ebaka, South Cameroon.

21.   The Leeukop and incipient dome, western Transvaal, R.S.A.

22.   Incipient dome, Halfway, near Johannesburg, R.S.A.

23.   Runnels extending into subsurface, Dumonte Rock, S.A.

24.   Contour plan of Yarwondutta Rock, S.A.

25.   The Humps and adjacent lateritised plain, W.A.

26.   Summit planation surface, Sermasoq, Greenland.

27.   Bornhardts in gneiss, Reynolds Range, N.T.

28 (a). Domes in sidewalls of Thompson River valley, Colorado.

   (b). Castle koppie, Rocky Mts high plain, Colorado.

29.   Castle koppie, Hoggar Mts, southern Algeria.

30.   Nubbin, western Pilbara, W.A.

31.   Blackingstone Rock, eastern Dartmoor, England.

32.   Castellated blocks on dome, Devil's Marbles, N.T.

33.   Angular blocks with tafoni on dome, Remarkable Rocks, Kangaroo Island,
            S.A.

34.   Plan of Gokomere inselberg, Zimbabwe.

35.   Inselberg, part granite, part gneiss, central Namaqualand, R.S.A.

36.   Sheeting structure, Mt Bundey, N.T.

37.   Nubbin with sheet structure exposed, western Pilbara, W.A.

38.   Inselberg, Paulshoek, Namaqualand, R.S.A.

39.   Low dome, Mt Bundey, N.T.

40.   Low domes, western Pilbara, W.A.

41.   Development of nubbins.

42.   Castle koppie, Devil's Marbles, N.T.

43.   Castle Rock, a castle koppie near Albany, W.A.

44.   Castle koppie in Tosa Gargantillar, Andorra.

45.   Haytor, eastern Dartmoor, England.

46.   Ranc de Bombe, Massif Central, France.

47.   Rooiberg, central Namaqualand, a bornhardt with ribbed slopes.

48.   Development of castle koppies.

49.   Morphological map and section of Devil's Marbles and environs, N.T.

CHAPTER 6.

1.   Section through Peruvian Andes.

2.   All-slopes in granite, south Greenland.

3.   All-slopes in granite, southeastern Brazil.

4.   Agulhas Negras, all-slopes in syenite, southeastern Brazil.

5.   Palaeosurface remnant, Andorran Pyrenees.

6.   'Gendarmes', Andorran Pyrenees.

7.   Sketch of Rio landscape.

8.   Sketch of butte, Ft. Lamy region, southern Sahara (Foureau).

9.   Faceted slopes in granite, northwest Q'd.

10.   All-slopes as etch surface, northern Flinders Ranges, S.A.

CHAPTER 7.

1.   Unconformity between granite and sandstone, Cape Town, R.S.A.

2.   Section through unconformity and exhumed surface, north Greenland.

3.   Stages in evolution of etch plain.

4.   Etch surface (New plateau), central W.A.

5.   Section through Waulkinna Hill and pediment, northern Eyre Peninsula,
            S.A.

6.   Inselberg and piedmont angle, Mauretania.

7 (a). Pediment 'fan', Namaqualand, R.S.A.

  (b). Coalesced pediment 'fans', Namaqualand, R.S.A.

8.   Topographic map of part of Eyre Peninsula, S.A.

9.   Topographic map of Cima Dome, California.

10.   Plan of Corrobinnie platform, northern Eyre Peninsula, S.A.

11.   Narrow basal platform, Pildappa Rock, Eyre Peninsula, S.A.

12.   Platform adjacent to Pildappa Rock, Eyre Peninsula, S.A.

13.   Map of Pildappa Rock and environs, Eyre Peninsula, S.A.

14.   Flared boulders and narrow platforms, Waulkinna pediment, northern Eyre
          Peninsula, S.A.

15.   Map of Lightburn Rocks, eastern Great Victoria Desert, S.A.

16.   Map of Ayers Rock and environs, N.T.

17.   Rock platform west of Ayers Rock.

18.   Peneplain on (a) northwestern, (b) northeastern, Eyre Peninsula, S.A.

19.   Inselberg and peneplain, western Cape, R.S.A.

20.   Alluvial plain, Georgina River valley, Q'd.

21.   Multicyclic landscape, Kamiesberge, Namaqualand, R.S.A.

22.   Valley side facets, Kamiesberge, Namaqualand, R.S.A.

23.   Map of stepped topography, southern Sierra Nevada, California.

24.   Development of stepped topography, according to Wahrhaftig.

25.   Development of stepped slopes by scarp foot weathering.

26.   Stepped slope of Poondana Rock, Eyre Peninsula, S.A.

CHAPTER 8

1.   Oblique aerial view of Pildappa Rock, S.A.

2.   Basins on part of upper surface of Pildappa Rock, S.A.

3.   Pan influenced by fractures, Johannesburg, R.S.A.

4 (a). Pit on Pildappa Rock, S.A.

   (b). Pan, Kulgera Hills, N.T.

   (c). Armchair-shaped hollows, Pildappa Rock, S.A.

   (d). Cylindrical hollow, Kwaterski Rocks, S.A.

5.   Distribution of rock basins, Pildappa Rock, S.A.

6.   Elongate pits along fracture, Ebaka, South Cameroon.

7.   Rock basins developed along fracture, Peella Rock, S.A.

8.   Elongate pit, Lightburn Rocks, S.A.

9.   Development of pit.

10.   Meringue surface, northwestern Eyre Peninsula, S.A.

11.   Pan on Yarwondutta Rock, S.A.

12.   Development of cylinder.

13 (a). Saucer-shaped depressions, Dumonte Rock, S.A.

    (b). Dimpled surface, Kwaterski Rocks, S.A.

14 (a). Pedestal rock, Domboshawa, Zimbabwe.

    (b). Pedestal rock, Ucontitchie Hill, S.A.

15.   Shallow depressions at base of flared slope, Hyden Rock, W.A.

16.   Moat at base of boulder, Tolmer Rock, S.A.

17.   Development of moat by drip and pool weathering.

18.   Rock doughnut, Enchanted Rock, Texas.

19.   Basal tafone and pedestal, Murphy's Haystacks, S.A.

20.   Gutters in bed of Ashburton River, W.A.

21.   Gutters on upper surface of Yarwondutta Rock, S.A.

22.   Remnants of gutters, Bruce Rock, W.A.

23.   Remnants of gutters, Caloote, S.A.

24.   Gutter extending below soil level, Pildappa Rock, S.A.

CHAPTER 9

1.   Flared basal slopes, Ucontitchie Hill, S.A.

2.   Flared boulders, Alabama Hills, California.

3.   Flared boulders, Bury Hills, Zimbabwe.

4.   Flared boulder, Sidobre, France.

5.   Flared slopes in arkose, Ayers Rock, N.T.

6.   Flared slopes, Pearson Islands, S.A.

7.   Wave Rock, Hyden, W.A.

8.   Slope with multiple flares, Pildappa Rock, S.A.

9.   Flared slope following hill-plain junction, Chilpuddie Hill, S.A.

10 (a). Overhanging flared slope on spur, Ucontitchie Hill, S.A.

   (b). Flared spur, Dinosaur, S.A.

11.   Flares in joint cleft, Yarwondutta Rock, S.A.

12 (a). Chinese hats, Nonning, northern Eyre Peninsula.

   (b). Mushroom rock, Sierra Nevada, California.

   (c). Mushroom rock, southern Libya.

   (d). Acuminate blade, Albany, W.A.

   (e). Anvil rock, Caloote, S.A.

13.   Stages in development of flared slopes.

14 (a). Yarwondutta Reservoir, S.A.

   (b). Excavation at Chilpuddie Hill, S.A.

15.   Flares merging with tafoni at Kokerbin Hill, W.A.

16.   Basal fretting at Poldinna Rock, S.A.

17.   Cumberland Stone, Scotland.

18.   Menhir, Dartmoor.

19.   Undercut basal slope at edge of ephemeral lake, Balladonia, W.A.

20.   Platform at the Humps, W.A.

21.   Scarp foot depressions at (a) Yarwondutta Rock, S.A., (b) Wattle Grove
            Rocks, S.A., and (c) Alice Springs area, N.T.
22.   Scarp foot depression, Mojave Desert, California.
23.   Scarp foot depression, eastern Egypt.
24.   Piedmont angle, Naraku, N.W. Q'd.
25.   Runnels, Singapore.
26.   Fluted boulder, Tampin, West Malaysia.
27.   Fluted block, Geelong, Victoria.
28.   Fluted overhanging slopes (a) Remarkable Rocks, Kangaroo Island, S.A.;
            (b) Ucontitchie Hill, S.A.; (c) Murphy's Haystacks, S.A.
29.   Pitted channel floor, Ucontitchie Hill, S.A.
30.   Fluted lower slope of bornhardt, western Cape, R.S.A.
31.   Boulder with fluted upper slope, Tampin, West Malaysia.

CHAPTER 10.

1.   Plan of Enchanted Rock cave, Texas.
2.   Interior of part of Enchanted Rock cave.
3.   Tafoni (a) Mt Hall, S.A.; (b) Hong Kong; (c) Ucontitchie Hill, S.A.;
            (d) Remarkable Rocks, Kangaroo Island, S.A.; (e) Pearson Islands,
            S.A.
4.   Tafoni related to fractures, Sardinia.
5.   Shelter in gritstone, Pennines, England.
6.   Alveoles, Namibia.
7.   Alveoles (a) Antarctica; (b) Hong Kong.
8 (a). Mamillated ceiling Ucontitchie Hill, S.A.
   (b). Ribbed ceiling, Tcharkuldu Hill, S.A.
9.   Flaking in tafoni, Ucontitchie Hill, S.A.
10.   Boulder tafoni, Kokerbin Hill, W.A.
11.   Tafoni and basal fretting, northern Transvaal.
12.   Scratch circles, South East of S.A.

CHAPTER 11

1.   Split boulder, Devil's Marbles, N.T.
2.   Split boulder, Tampin, West Malaysia.
3.   Stages in development of split boulder (a) Devil's Marbles, N.T.;
            (b) Massif Central, France; (c) Palmer, S.A.
4.   Suggested mode of development of split rock.

5.   Split blocks, Kokerbin Hill, W.A.

6.   Parted block, Dartmoor.

7.   Complex parted blocks (a) Devil's Marbles, N.T.; (b) Dartmoor.

8.   Polygonal cracks, Tcharkuldu Hill, S.A.

9.   Heiroglyphs, Daadening Hill, W.A.

10. Polygonal cracks on platform, Corrobinnie Hill, S.A.

11. Star fractures, Corrobinnie Hill, S.A.

12. Multiple shells with polygonal cracks, Mt Magnet, W.A.

13. Polygonal cracks, northern Transvaal, R.S.A.

14. Corestones with polygonal cracks, Snowy Mts, N.S.W.

15. Development of polygonal cracks by precipitation of salts along partings.

16. A-tent, Mt Wudinna, S.A.

17. A-tent, Enchanted Rock, Texas.

18. Thick A-tent, Mt Wudinna, S.A.

19. Plans and sections of A-tents.

20. Thin A-tent, Carappee Hill, S.A.

21. Overlapping slab, Mt Wudinna, S.A.

22. Section through overlapping slab.

23. Displaced slab, Little Wudinna Hill, S.A.

24. Plan of displaced slab, Little Wudinna, S.A.

25. Slipped slab, Mt Wudinna, S.A.

26. Residual blocks, Pildappa Rock, S.A.

27. Slabs displaced by tree roots, Mt Wudinna, S.A.

28. Vertical wedge, Mt Wudinna, S.A.

29. Plan of, and sections through, vertical wedge.

30. Stages in development of arched and angular tents.

PREFACE

   Granite landforms and landscapes have long engaged the interest and attention
of geologists and geomorphologists. Yet, so far as I am aware, no one has
attempted an analysis and synthesis of granite forms and their genesis presented
in the English language. Nothing has been done for the morphology of granite
exposures that can reasonably be compared to Marjorie Sweeting's *Karst
Landforms* (1972). Wilhelmy (1958) contributed an excellent account in German
and organised in terms of climatic zonation (see also Wilhelmy, 1974), though
in this work it is significant that many arguments are proposed only so that
they can be refuted either in whole or in part. Alain Godard (1977) produced
a marvellously concise review of granite landforms and landscapes in the French
language. Certain granite forms are discussed at some length in Thomas'
(1974a) exposition concerned with tropical geomorphology, but the self-imposed
climatic delimitation of interest places obvious and critical limitations on
this work seen in the context of the analysis of granite forms. In addition to
these major publications specific features and problems associated with outcrops of granite have been the subject of splendid essays and papers, and many
are cited in this book. Two examples may be cited to illustrate their virtues
but also their inherent limitations in the overall context. Dale's (1923)
account of the granites of New England contains many perspicacious and seminal
geological and geomorphological observations that in many respects remain
unsurpassed. His speculations on the origin of sheet structure for instance
are extraordinarily astute, though incidental to his main theme, which was the
commercial value of the granites. Again, the recent *Supplementband* on inselbergs (Bremer and Jennings, 1978) is concerned specifically with a particular
and well-known granite form, but does not purport to be a comprehensive review
of all aspects of even that single feature; rather it is a collection of
essays on topics that happen to interest the contributors.
   There is therefore a perceived need for a systematic, coherent and comprehensive account and analysis of granite landforms. Additionally, however, in
browsing through the literature it is apparent that several of the concepts
that currently find favour in the interpretation of granite forms and that are
regarded as of modern derivation were in fact appreciated and applied long ago,
and in some instances more than a century ago. Here an attempt has been made
to trace the origins of explanations of various granite landforms, both major
and minor, in an endeavour to give credit where it is due. Last, though it
cannot be claimed that any of the explanations propounded here is new, it is

nevertheless suggested that certain ideas warrant restatement and fresh emphasis, and that their implications require re-examination in light of a wider field experience.

The book is as uncomplicated in its organisation as it is simple in concept. In order to arrive at an explanatory account of the forms developed on granitic rocks it is necessary first to define and describe those features that are to be considered, together with the nature of the materials on which they have evolved and the weathering processes at work at and near the earth's surface. These topics, together with a brief anticipatory review of major themes constitute the three chapters that are Part I. Major landforms and assemblages are discussed in the four chapters of Part II and the minor features that have evolved on the major hosts provide the material of the four chapters of Part III. Part IV comprises but a single chapter in which salient conclusions are discussed.

Over the years many friends and colleagues have contributed in one way or another to the writing of this book, from O.D. Kendall, F.C. Phillips and Frank Hannell who many years ago introduced me to the tors and high plains of southwestern England, to those who have at various times over the past several years guided me, or discussed landscapes and ideas in the field: 'Po Po' Wong in Singapore, and a former student, Aloyah Salleh, in West Malaysia; Bill Bradley, Charles Higgins, Clyde Wahrhaftig, Bob Folk, Craig Kochel and Ernst Kastning in the United States; Rodney Maude, Lester King, Linley Lister, Roy Miller, Bill Purves, Bernie Moon, Tim Hill, Eban Verster, Braam de Villiers, 'Flip' Hattingh and Koos van Zyl in southern Africa; Regina de Meis in Brazil; Pierre Barrère, Serge Morin, Marie Claire Prat and Max Durruau in France; and in Australia Rudi Horwitz, Joe Jennings, Jennie Bourne, George Sved, Jan de Jong, Tony Milnes and Edwin Hills.

I am also grateful to those many friends, colleagues and institutions who have allowed me to use their photographs and illustrations. Many of these sources are named in the text, but particular thanks are due to those who at my request sent photographs of admirable quality and interest, but for which no space could, in the event, be found in the text. I am as indebted to these unnamed friends as I am to those whose assistance is specifically acknowledged.

Max Foale, Chris Crothers and Debbie Oakley displayed patience and ingenuity in preparing the line drawings; Anne Whicker typed the final camera-ready sheets with speed and style; and my wife Kate typed and effectively edited the final draft of what was, regrettably, a fairly messy manuscript. Ron Gramende and his colleagues at Elsevier have at all times offered patient understanding and helpful advice in the face not only of the usual sorts of problems that afflict authors and publishers but also those arising from the author's living in the most striking country in the world, Australia.

This book is based not only on the distilled essence of the experience and wisdom of many earlier workers, but also in quite extensive field experience in many parts of the world. Some of these field investigations which form a crucial basis for any study of this type, have been made possible by grants from the Australian Research Grants Committee, and from study leave funding from the University of Adelaide. In earlier years a Nuffield Commonwealth Bursary and an U.S. National Science Foundation Fellowship materially assisted in field studies outside Australia. To all of the people, institutions and organisations who have in any way contributed or assisted in any way I offer my sincere thanks.

C.R. Twidale
Adelaide, April, 1982.

PART I

INTRODUCTION

CHAPTER 1

LANDFORMS DEVELOPED ON GRANITIC ROCKS

Granite has not one face but many. Not only has the rock been shaped into
uplands in some places and worn down to extensive plains of low relief elsewhere, but the form of these hills and plains varies greatly from place to
place. Granitic masses have been exposed to the entire range of endogenetic
and exogenetic forces and agencies, in large measure with the same results as
in other rock types. None of the landforms developed on granitic rocks is
peculiar to those materials, for congeners of all major and most, perhaps all,
minor forms can be found developed in other lithological environments.

Granites are subjected to normal crustal processes and are, for example,
affected by faulting. Fault scarps, horsts and rift valleys are developed in
granitic terrains as in regions dominated by other rock types (Fig. 1.1).

Fig. 1.1. The Sierra Nevada of California is delineated on its eastern side by
a complex fault zone that trends roughly north-south. Near Lone Pine a major
fracture zone defines the upland proper and finds topographic expression in a
dissected fault scarp about 1,500 m high. The granitic rocks of the Sierra
have here been deeply eroded by rivers and shattered by frost action, producing
an all-slopes topography. The acicular peaks of the region include Mt Whitney,
at 1,347 m the highest peak in the mainland United States outside Alaska. A
lower downfaulted block, also granitic, gives rise to low boulder-strewn rises
known as the Alabama Hills and much beloved by the makers of Western films:
many a "shoot-out" has been staged amongst the boulders of this foothill region.

Faulting has caused the displacement of outcrops and the landforms developed on them (Fig. 1.2). Faulting has brought into juxtaposition rocks of contrasted character and, for this reason, granitic terrains in some areas give way abruptly and along dramatically linear boundaries to quite different landscapes (Fig. 1.3). Fault zones are exploited by weathering and erosion and may produce fault-line valleys that are characteristically straight. Thus the direct and obvious major effects of faulting in granite are not significantly different from those in other lithological environments and warrant no further comment here.

Outcrops of granite are affected by every geomorphological process known to operate at and near the earth's surface, and many of the assemblages so developed are indistinguishable from those moulded by the same processes in other rock types. Thus as is illustrated in various subsequent chapters glaciated granitic masses in such areas as southern Greenland and the Yosemite region of the Sierra Nevada of California are characterised by cirques, tarns and troughs, by hanging valleys, valley steps and *roches moutonnées*. Certainly domical forms are prominent in such areas, but the general suite of forms is typical of glaciation rather than of granite bedrock (Fig. 1.4). Similarly, river channels cut in granite display pot holes, scallops and flutings, as well as structurally controlled rapids and waterfalls. In certain, rather particular, circumstances, and both in warm and in cold climates (see e.g. Russell, 1932; Powers, 1936), aeolian sand blasting has caused granitic rocks to become intricately fluted and scalloped, but again there are analogues in other lithological settings.

In coastal contexts, too, the gross assemblage of forms is due to the processes operating there and not to properties peculiar to granites. Tafoni and flutings are particularly well-developed (Chapters 9 and 10), and granite domes and associated sheet structure are prominent in some localities (Fig. 1.5). Orthogonal fracture sets also find marked expression (Fig. 1.6) but, with few exceptions, granite coasts are much the same as most others.

It is true that corestones and grus (see Chapter 4) are exposed in some granite cliffs and, on being released by the erosion of the grus, the corestones form boulder beaches, some of which display fretted, fitted and imprisoned boulders and blocks (Hills, 1963). The principal oddity of granite coasts however is a negative one for, as Jutson (1940) and Hills (1949, 1971) have pointed out, shore platforms are generally not well developed, if they are present at all, in fresh granite. Blocks and boulders are characteristic of well-jointed granite exposed at the coast, yet there are quite broad and extensive platforms underlain by fresh granite and extending in places fully 200 m from the base of the cliffs (Fig. 1.7) on the west coast of Eyre Peninsula (Twidale et al., 1977), and other, less extensive, granite platforms are developed elsewhere along the South Australian coast. The most extensive granite platforms have a rather

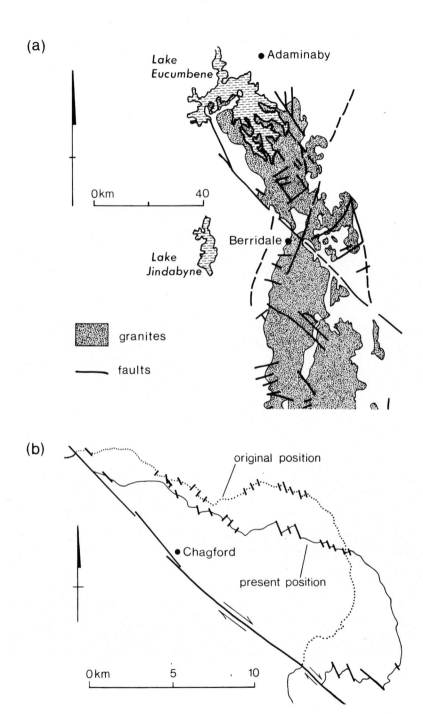

Fig. 1.2. Displacement of granite masses by faulting: above (a) the Berridale Batholith, N.S.W. (after Williams *et al.*, 1975); below (b) on eastern Dartmoor, southwestern England (after Blyth, 1962).

4

Fig. 1.3. The MacDonald Fault Scarp, a fault-line scarp which in places attains a height of 280 m, has been traced for more than 500 km in the North West Territories of Canada. Older granitic rocks underlie the higher ground to the left of the scarp, and sediments of the Et-then Series, also of Precambrian age, the lower ground on the downthrow side of the fault (Dept. Energy, Mines, Resources, Canada).

Fig. 1.4. Half Dome, seen here from upper Yosemite Falls, Sierra Nevada, California has been affected by glaciers but the gross morphology is an expression of structure (C. Wahrhaftig).

unusual origin. Waves operating within the present tidal zone have cut into an old regolith developed on Precambrian crystalline rocks and stripped off the unconsolidated material to expose the essentially planate and horizontal weathering front as an etch surface which, since it occurs in the coastal zone, is called a platform. The boulders that are strewn over many of the platforms are erstwhile corestones.

But having directed attention to similarities between the landform assemblages shaped in granite, and those developed in other rock types, it must also be stated that many granitic landscapes are nevertheless distinctive, and that many individual landforms are better and more widely developed on granite than in other lithological environments.

Major forms characteristic of granitic exposures are boulders, inselbergs, all-slopes topography and plains. Possibly the most common and widespread

6

Fig. 1.5. The Pearson Islands are members of the Investigator Group, situated some 65 km off the west coast of Eyre Peninsula, S.A. Note the cliffed western (right) face of the islands exposed to wave attack from the Great Australian Bight (Keith P. Phillips).

granite landform is the boulder - a more-or-less rounded mass standing either in isolation or in groups or clusters (Fig. 1.8) on plains, in valley floors, on hillslopes or on the crests of hills. Granite boulders have been described from all climatic zones, from polar to equatorial, and from arid to humid. They vary in degree of roundness, and in diameter from 25 cm to some 33 m, though the mode is of the order of 1-2 m.

Some granite uplands take the form of ranges or massifs; some of them are detached, whilst others form integral parts of larger mountain ranges. Thus the Everard Range is a complex of granite hills standing in isolation above the arid plains of central Australia (Fig. 1.9); the uplands of central Namaqualand consist of an orderly series of juxtaposed fracture-controlled blocks (Fig.1.10); and the Spitzkoppe comprise a connected series of acicular hills and domes rising from the plains of central Namibia (Fig. 1.11). On the other hand, the rugged granite hills of the Mt Painter area are part of the northern Flinders Ranges; the spectacular domical forms of the Yosemite and Domeland are developed within

Fig. 1.6. Orthogonal joint sets exposed in coastal cliffs at Land's End, south-western England (Geol. Surv. and Museum, London).

the wider context of the Sierra Nevada in California; and Corcovado, for example is part of the Tijuca Massif of the Rio de Janeiro area.

Other granite hills stand in isolation. This class includes some of the best known granite eminences. Lone hills have a special dramatic quality, rearing abruptly as they do from the surrounding plains or low hills: these are the hills known as *Inselberge* (English plural, inselbergs) or island mountains (German: *Insel* - island; *Berg* - mountain) because of the abrupt transition from plain to upland, and the steep flanks of the uplands that are reminiscent of sea cliffs. Describing the *Inselberglandschaften* of central Australia, Giles (1889, p. 158) compared such residual hills to islands, and Bornhardt (1900), Holmes (1918) and Passarge (1895) employed the same simile. The aptness of the analogy is brought home by the brilliant field sketches (Fig. 1.12) pro-duced by Bornhardt (1900), Passarge (1928) and Jessen (1936), and by a comparison of typical inselberg landscapes (Figs. 1.13-16) with examples of offshore insel-bergs in such areas as the Bahia Guanabara (essentially Rio Harbour - Fig. 1.16); Wilson Promontory, Encounter Bay, the Investigator Group (Fig. 1.5), Esperance

Fig. 1.7. On the west coast of Eyre Peninsula the unconformity between Pre-cambrian rocks (commonly granite, but also including gneisses) and Pleistocene dune calcarenite frequently occurs within the present tidal and spray range. As a result the weaker limestone has in places been stripped away, as has the grus developed at the granite surface. The exposed weathering front, together with the boulders (corestones) released by the evacuation of the grus, now forms shore platforms strewn with boulders and of a width unusual for granitic environments. This example occurs just south of Streaky Bay.

Bay and King George Sound, all in southern Australia; and in southwestern England where the Scilly Islands and Lundy are literal as well as littoral island mountains.

Whether they are extensive or of limited area, whether they are simple or com-plex, whether they are uniform or varied, these granite uplands are susceptible of rational analysis, for granite uplands appear to comprise repetitions and mixtures of a few basic forms, examples of which are shown in Figs. 1.13-18.

Bornhardts, so named by Willis (1934, 1936) after the German traveller and explorer to whom are due the first scientific descriptions of the forms (Bornhardt, 1900), are bald, steep-sided domes. They vary in size and geometry. Those that are low, distinctly elongate and elliptical in plan, and have very steep sides are called whalebacks (*dos de baleine*); some that are more nearly symmetrical in plan, and have largely steeply sloping flanks, are known as turtlebacks; those high residuals that are asymmetrical in profile are called elephant rocks, or *dos d'éléphant* (see e.g. Rognon, 1967); and yet others that

Fig. 1.8. (a), Above, large single granite boulder, some 4 m diameter, near the Herbert Falls, north Q'd (Div. Land Res. C.S.I.R.O.); (b) below, cluster of granite boulders, near Palmer, eastern Mount Lofty Ranges, S.A.

have plan axes of similar length, roughly equal to the altitude of the hill above the plains, are referred to simply as domes or half-oranges, though there are many local names that are used in the same general sense (*matopos, ruwares, morros,*

Fig. 1.8. (c) Granite boulder near Tampin, West Malaysia, where the slopes have been cleared of rain forest in preparation for the planting of rubber trees. The people on the path (bottom) provide a scale.

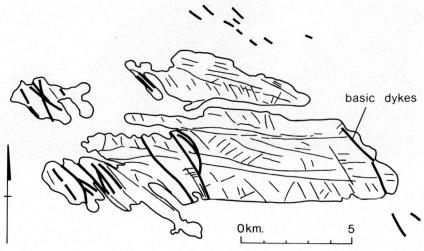

basic dykes

0km.   5

Fig. 1.9. Map of Everard Range, northern S.A., showing major fractures and dolerite intrusions (Drawn from air photographs, Sprigg *et al.*, 1959; Krieg, 1972).

Fig. 1.10. The Groothoekseberg and adjacent bornhardts, in the Kamiesberge of central Namaqualand, western Cape Province, R.S.A.

Fig. 1.11. The Groot Spitzkoppe is a group of acuminate gneissic inselbergs in central Namibia, surrounded by broadly rolling gibber-veneered plains in the foreground and middle distance.

Fig. 1.12. An example of the sketches of Angolan inselberg landscapes drawn by Jessen.

Fig. 1.13. Bornhardts of augen gneiss rising from plains veneered by sand but underlain by gneiss at a depth of a few centimetres in the western Pilbara, near Nanutarra, W.A.

Fig. 1.14. Isolated granitic inselberg standing above the extraordinarily flat Bushman Surface, also cut in granite, in central Namaqualand, R.S.A.

Fig. 1.15. Granite domes and peaks in the Sahara Desert, southern Algeria (P. Rognon).

Fig. 1.16. The Sugar Loaf (*Pão de Assuçar*, or *Açuçar*) and other residuals bordering Rio Harbour in the humid subtropical coastlands of Brazil (Braz. Tour. Bur.).

Fig. 1.17. A bouldery nubbin surrounded by very gently sloping ($\frac{1}{2}^0$ - $2\frac{1}{2}^0$ inclination from the horizontal), mantled pediments near Naraku, northwest Queensland. The thickness of granite sand on the pediment varies between nil or a centimetre near the hill base to about 50 cm some 200 m distant. Termite mounds have been built, despite the thinness of the regolith.

Fig. 1.18. This castellated inselberg or castle koppie in the Mrewe-Marandellas area of Zimbabwe is built of huge cubic and quadrangular blocks *in situ*.

Fig. 1.19. Domes and turrets in the Kilba Hills, northern Nigeria (drawn from photograph in Falconer, 1911).

dwalas, meias laranjas, demi-oranges, and so on) and which, from their linguistic derivations, go some way to indicating the wide distribution of the forms. Forms that are high and narrow, at least along one plan axis, are the sugar-loaves and turrets of, for example, the Rio de Janeiro area (Fig. 1.16), the Djanet region of the Tassili and parts of Nigeria (Fig. 1.19). Some hills take the form of large radius domes, others small; some stand high above the adjacent plains, others are little more than low platforms; some are extensive, others are of limited areal extent; some are simple domes, some carry blocks on their crests, and yet others are stepped. But whatever their precise geo-metry, all are bald, steep-sided domical hills and are bornhardts.

Some block- or boulder-strewn hill-sized inselbergs are called *nubbins* or knolls (Fig. 1.17). Angular and blocky, castellated forms (Fig. 1.18) are known as *castle koppies* (or *kopjes; koppie* is Afrikaans for a head, is colloquially applied to a small hill and is preferred). Many inselbergs contain elements of two or even of all of these types, though it is thought that both koppies and nubbins are derived from bornhardts, which are regarded as the basic form (Twidale, 1981a).

Domical forms frequently occur in ordered groups separated by fracture-controlled valleys or clefts (Figs. 1.9 and 1.10). In such instances the indi-vidual domes can be referred to as bornhardts, but they cannot be termed inselbergs because they do not stand in isolation. However, the group considered as an entity does, and hence is an inselberg. Thus, though there are many bornhardts within the Everard Range, it is the range itself that comprises the inselberg. Ordered assemblages of castellated forms are rare, though a few examples are known. Nubbins, on the other hand, are commonly grouped, as for instance in the Mt Bundey area, near Darwin, N.T., in the western Pilbara of Western Australia, and near Alice Springs, N.T.

Reference has already been made to terms of local derivation that are applied to bornhardts. Some such terms are troublesome because they have come to be used in different senses in different parts of the world. In particular, the word 'tor' (Twidale, 1971a; pp. 14-17) is used in England of angular residuals, most of them granitic (Linton, 1952, 1955), but some developed in gritstone (Palmer and Radley, 1961; Linton, 1964) or in dolomite (Ford, 1962). Tor was used in this sense by Falconer (1911) for residuals in Nigeria, by Handley (1952) to denote granitic hills in Tanganyika, and by Demek (1964a and b) and Jahn (1974) of castellated forms in the Bohemian Massif and Karkonosze Mountains of southern Poland. In recent years, however, the term has been used with a particular genetic connotation in mind (Linton, 1955, 1964 - see Chapter 5). Not only has the particular interpretation been challenged (see e.g. Palmer and Nielson, 1962), but in some parts of the world, particularly in Australasia (Jutson, 1914, p. 182; Williams, 1936; Hills, 1940, pp. 26-28; Cotton, 1941, pp. 28-30; Costin, 1950; Sparrow, 1961; Browne, 1964), the term has been used of isolated blocks and boulders*, a sense in which it was also sometimes applied in southwestern England (see e.g. MacCulloch, 1814). Boulders in Namibia (Mabbutt, 1952) and in Nigeria (Thomas, 1965) have also been described as tors and Mabbutt (1965) has applied the term to a small bornhardt in central Australia; in other instances, however, it is difficult to know what form is implied.

Tor is a word of ancient derivation. It is Cornish and is comparable to the Anglo-Saxon *torr,* the Welsh *twr,* and the Latin *turris* meaning a tower. It has long been used of protuberant rock outcrops. Jones (1859, p. 132), for instance, described the tors of Dartmoor as 'Irregular prominent masses of rudely heaped rock fragments', which can be construed as indicating nubbins, but Linton (1952, p. 354) formally defined a tor as a bare rock outcrop, usually of monumental form, 'about the size of a house' (and whether a labourer's cottage or a ducal palace, of some considerable size), commonly bounded by near-vertical fractures and boldly fissured by widely-spaced joints. In view of Linton's interest in Devon and Cornwall, where the most common granite residuals are castellated, it

---

* Professor E.S. Hills (pers. comm., 7 December, 1981) relates that when he was a boy in Melbourne the word 'tor' meant a marble or alley, albeit of a special sort - of tough agate rather than of common glass. Clearly the morphological analogue was the rounded boulder. Yet in a geomorphological sense 'tor' was and is applied in the Antipodes to all manner of granite residuals, including born-hardts (see e.g. Noldart and Wyatt, 1962, p. 51).

is not without significance that of the three main types of inselberg this
definition comes closer to that of the koppie forms than either of the other two,
though the residuals that constitute the 'houses' in Anderson's (1931) Cassia
City of Rocks, in Idaho, are simply very large, rounded residual boulders.

Whatever the rationale, the term tor has been usurped by workers outside
Britain and even in its home country was used in a different sense in earlier
times. Features similar to the tors of southwestern England occur in different
and various climates in other parts of the world are commonly called inselbergs
and in particular castle koppies. If the tors of England occurred anywhere else
they would be referred to as inselbergs and most of those of Dartmoor, for
example, as castle koppies.

Another difficulty is that the rounded or subrounded masses known as tors in
other parts of the world, and especially in Australasia, are really boulders in
the usage of sedimentary petrologists. Thus a boulder is defined as 'a detached
rock mass, somewhat rounded or modified by abrasion in transport' (Lane *et al.*,
1947) or by weathering *in situ*, and 'larger than a cobble' (Pettijohn, 1957,
p. 20). The minimum diameter of a boulder is set at 256 mm, or about ten inches
(Lane *et al.*, 1947),with no upper limit. This definition embraces all the
boulder tors described in the literature and it is interesting to note in passing that the word boulder was used in some early accounts of the forms and that
it has never really lost currency (see Chapter 4).

Some granite uplands are neither domical nor castellated nor block- or
boulder-strewn. Instead, the granite massifs have been deeply dissected, slopes
are essentially rectilinear and there are no significant areas of flat land
either in valley floors or on ridge crests. All-slopes topography (Fig. 1.20)
has developed on granitic rocks in such arid areas as the Sinai Peninsula and
the northern Flinders Ranges, in cold areas such as the Andes, the Karkonosze
Mountains of southern Poland, the Pyrenees and the eastern Sierra Nevada (e.g.
the Mt Whitney area - see Fig. 1.1), and in such humid subtropical areas as the
Serra do Araras, near Rio de Janeiro.

Commonly associated with these major forms is a suite of minor features -
sheet structure, rock basins, flutings, flared slopes, tafoni (Figs. 1.21-24)
and so on - features which, though neither ubiquitous nor peculiar to granite,
are nevertheless widely and well developed on that material.

Notable and distinctive as these upstanding forms undoubtedly are, however,
plains are, on an areal basis, far more representative of granite outcrops.
Although there are many extensive granite uplands (the Kamiesberge of Namaqualand,
in the western Cape Province, R.S.A., considerable areas of the Sierra Nevada,
the Serra da Mantiqueira of southeastern Brazil, and so on), there are even more
extensive granite plains broken only by a few spectacular but areally insignificant residual hills (Fig. 1.14).

18

Fig. 1.20. All-slopes topography in the northern Flinders Ranges, S.A. (S.A. Tour. Bur.).

Fig. 1.21. Sheet structure exposed in coastal cliffs, Pearson Island, S.A.

Fig. 1.22. A flat-floored rock basin or pan on Haytor, eastern Dartmoor, south-western England.

Fig. 1.23. Fluted and flared slope on the northwestern margin of Pildappa Rock, northwestern Eyre Peninsula, S.A.

Fig. 1.24. Tafoni developed beneath and at the edge of a massive sheet structure on Ucontitchie Hill, northwestern Eyre Peninsula, S.A.

These granite plains take at least four forms. Pediments carrying a veneer of grus or weathered granite *in situ*, and sloping gently and smoothly down from residual uplands, are well and widely developed, particularly in tropical and subtropical lands. Such mantled fringing pediments on granite occur in northern Algeria, the American Southwest, Namibia, northwestern Queensland, central Australia, Eyre Peninsula, Chile, Japan and Korea, to name but a few localities where they are well represented (Figs. 1.13, 1.17 and 1.25; see Whitaker, 1973, and Twidale, 1981b, for references).

Rock pediments or platforms are, like mantled pediments, most commonly found adjacent to uplands, but they also occur on hillslopes, in valley floors and on hill crests lacking residual remnants. They lack a debris veneer (Fig. 1.26), and consist of essentially flat or gently sloping rock platforms, all of which are, to a greater or lesser degree, grooved and dimpled through the presence of gutters and shallow saucer-shaped depressions (Twidale, 1978a; see also Chapter 8).

Both mantled and rock pediments are of comparatively limited area. The former extend for a few kilometres at most from the mountain front and most of the latter are only a few tens of metres wide, though some few in wet sites on

Fig. 1.25. Ucontitchie Hill, northwestern Eyre Peninsula, S.A., is a bornhardt surrounded by a low angle cone or pediment which is for the main part mantled, though bare rock platforms are exposed in places and the regolith is nowhere very thick in the vicinity of the upland.

Fig. 1.26. Bare rock pediment or platform bordering Corrobinnie Hill, northern Eyre Peninsula, S.A. and extending 700-800 m from the base of the upland. Note the shallow runnels and depressions.

northern Eyre Peninsula and in the southwest of Western Australia are a few
scores or even, in exceptional cases, a few hundreds of metres across.

Plains of great areal extent are developed on granitic rocks in many parts of
southern Africa and Australia. Some, like those just north of Pretoria, on
northern Eyre Peninsula, and in southwest Western Australia (Fig. 1.27) are
broadly rolling and can descriptively be called peneplains (Davis, 1909).
Others, like the Bushman Surface of Namaqualand (Fig. 1.14) and the plains
around Meekatharra, Western Australia (Fig. 1.28), are extraordinarily flat.
They have been called pediplains by some workers (e.g. King, 1942, p.53) because
it has been assumed that they result from the elimination of inselbergs and
coalescence of pediments; but it is an assumption, and they are perhaps best
referred to simply as planation surfaces or as ultiplains.

Thus four major landform assemblages - boulders, inselbergs, all-slopes
topography and plains - are widely and well developed on granitic rocks. Their
possible origins are discussed in the following chapters, partly for their
intrinsic interest, partly for the light they shed on general concepts of

Fig. 1.27. Peneplain cut in weathered granite near Corrigin in the southwest
of W.A.

Fig. 1.28. This extraordinarily flat plain is cut in fresh Archaean migmatite near Meekatharra, W.A.

landform development. But in order that the evidence and argument presented can be seen in context, three or four basic conclusions can usefully be anticipated by stating them in outline here.

Although many granite landforms have been discussed in structural terms (e.g. de Martonne, 1951; Twidale, 1971a), climatic interpretations involving the zonation of forms have also enjoyed considerable favour (Wilhelmy, 1958, 1974; Büdel, 1977). Indeed, many granite forms have been used as climatic indicators. Thus granite corestones and boulders were for many years interpreted as glacial erratics (Agassiz, 1865; Romanes, 1912) and low latitude glaciation was postulated on that basis. Flutings also were associated with glaciation during the early years of the Nineteenth Century, and Logan (1849) did not conceal his pleasure at finding examples on Singapore Island that he, quite rightly, thought refuted this then established idea. Inselbergs are most commonly treated in the context of the arid erosion cycle, or some similar heading. There are exceptions (e.g. de Martonne, 1925, 1951), but most geological and geomorphological texts that mention inselbergs treat them as desert or savanna forms. Thus, to take recently published works, Thomas (1974a) is slightly equivocal but leans toward a savanna setting, while Bloom (1978, p. 324) states that 'General agreement has now been reached that inselbergs are truly the result of savanna morphogenetic processes ...'.

It is not denied that certain forms developed on granitic rocks are more frequently found in some conventionally defined climatic zones than others, but many are widely distributed and others are limited to what appear to be two or more climatically disparate settings. Thus pediments are found well and widely developed in arid and semiarid lower latitude regions but occur also in subhumid and in cold climates. Tafoni are found in desert as well as coastal settings, and so on. But rock basins, boulders, inselbergs, and many other forms are widely developed and preserved.

This azonal character is attributed, first, to various structural factors that have had a significant influence in landform development on granitic rocks. The consequences of subtle variations in composition and texture within granitic masses have been neglected, but the effects of contrasts in fracture density are well-recognised. The reasons for such variations in density and associated stress conditions are also of interest. But, whatever the details, such structural factors function independently of climate and impose a measure of azonality on granite forms.

Second, although granite is stable in dry environments, contact with moisture leads to rapid and pronounced decay. The weathering of granite is almost wholly due to moisture attack. Subterranean waters in the form of either soil moisture or of groundwater are practically ubiquitous. Some of this subsurface moisture is derived from modern precipitation (for no place on earth, not even the driest desert, is rainless), but many groundwaters are of considerable antiquity and originated from earlier subsurface infiltration. Furthermore, it is increasingly apparent that through much of the Mesozoic and for all but the last two million or so years of the Cainozoic, warm humid conditions obtained over wide areas of some continents, even in high latitudes. Thus subterranean water is and has been widely distributed. Many of the landforms, major and minor, characteristic of granite outcrops were initiated by differential weathering at the lower limit of weathering, or weathering front. The latter is especially well-defined on granitic rocks because of the typical low permeability of the fresh rock. The climate at the land surface is largely irrelevant, save insofar as water temperature and throughflow influence the rate of weathering.

Thus many granite forms are in disequilibrium with the present climates. The flared slopes of southern Australia, for example (see Chapter 9), seem to be especially well developed in alkaline environments, but the carapace of calcrete that blankets the plains bordering the flared slopes is a late Cainozoic development and is largely irrelevant to the conditions in which the flared forms were initiated. Moreover, whatever agency - rivers, glacier ice, wind-driven waves - is responsible for evacuating the weathered mantle and exposing the irregular weathering front, the latter displays the same morphological range, both in gross and in detail, the world over. To pursue the example of flared slopes, the form is essentially the same whether exposed by rivers, waves or the wind (Figs. 1.23 and 1.29). This, then, is another reason for the observed azonality of granite forms.

Fig. 1.29. Flared boulder (F) on shore platform at Smooth Pool, near Streaky Bay, Eyre Peninsula, S.A. (Fig. 1.7). The flared shape was initiated by sub-surface moisture attack, and was exposed by wave erosion but is morphologically similar to that exposed by river work on the flanks of Pildappa Rock (Fig. 1.23).

These themes involving the interplay of subsurface weathering and structure in aeons long since past are detailed and elaborated in various of the following chapters. First, however, it is necessary briefly to review certain aspects of the occurrences and characteristics of granite as a rock, and of the changes that take place when granite is exposed to the atmosphere and to groundwaters, before embarking on discussions of major granite forms and then of the minor features that have evolved on the host masses.

CHAPTER 2

CHARACTERISTICS OF GRANITIC ROCKS

Granite is one of the best known of all rocks. It can be a beautiful orna-
mental stone and, rough cut or polished, many varieties are widely used as a
facing on public buildings and for monumental purposes. It is a measure of the
familiarity Everyman feels for granite that many other crystalline rocks are
frequently referred to as granite. The norite from Black Hill, South Australia,
for instance, is commonly known as 'black granite'.

A. OCCURRENCES OF GRANITE

Granitic rocks, including granite gneisses, are exposed over about 15% of the
continents, of which, therefore, they are a significant component. The study of
pressure waves generated by earthquakes shows that the earth is not a homogeneous
mass, but rather that it consists of shells of different materials surrounding a
core (Fig. 2.1 (a)). The innermost contiguous layer is composed of dense
material of basaltic composition. Known as the *sima*, a mnemonic derived from its
two principal components, *silica* and *magnesia*, it forms the floors of the oceans.
Floating in the sima are rafts of less dense *sial* (*silica* and *alumina*) which form

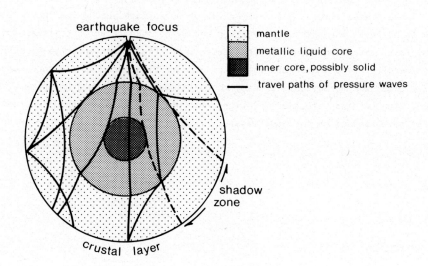

Fig. 2.1 (a). Structure of the earth (after Hodgson, 1964)

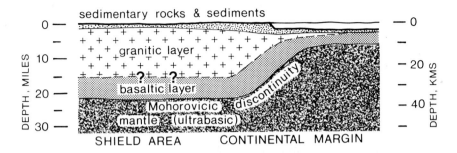

Fig. 2.1 (b). Structure of the upper crust with sial and sima (after Hodgson, 1964).

the outer discontinuous layer of the crust (Fig. 2.1 (b)). The sialic rafts are, of course, the continents. They are of granitic composition and occupy about one third of the earth's surface. Thin veneers of sedimentary rocks cover large areas of both the continents and ocean floors. Considerable areas of basaltic rocks also occur within the continental regions but in broad view there is a clear and significant distinction between the basalt of the deep ocean basins and the granitic continental masses.

Granitic rocks are a major component of the ancient crystalline blocks or shields that form the nuclei of all the continents (Fig. 2.2). These, the oldest parts of the earth's crust, comprise zones of different ages (Fig. 2.3 (a) and (b)) that have been welded one to another through time. The shields also appear to be areas of recurrent epeirogenic uplift. The crystalline rocks of the shield areas crop out over extensive areas of the continents but also under-lie many of the contiguous areas of essentially undeformed sediments, as, for example, in the Ukraine. Where the sedimentary cover thins sufficiently, out-crops of granitic rocks give rise to small, but frequently notable, inliers and morphological enclaves, as, for example, Bald Rock and Pyramid Hill, granite inselbergs standing above the sedimentary plains of northern Victoria. Granitic rocks are also a significant constituent of most fold mountains or orogenic belts (Fig. 2.4).

Granitic rocks vary in origin. In general terms, the character of the margins of the crystalline masses, considerations of space, and the detailed composition, texture and structure of the masses suggest that most of the very large masses of granitic rocks exposed in the shield areas are of metamorphic origin, having been formed by the alteration of pre-existing rocks by thermal and deformational events repeated throughout great spans of geologic times (see e.g. Read, 1957). On the other hand, many of the small bodies that are particularly characteristic of orogens are of magmatic origin in that they crystallised out from molten

28

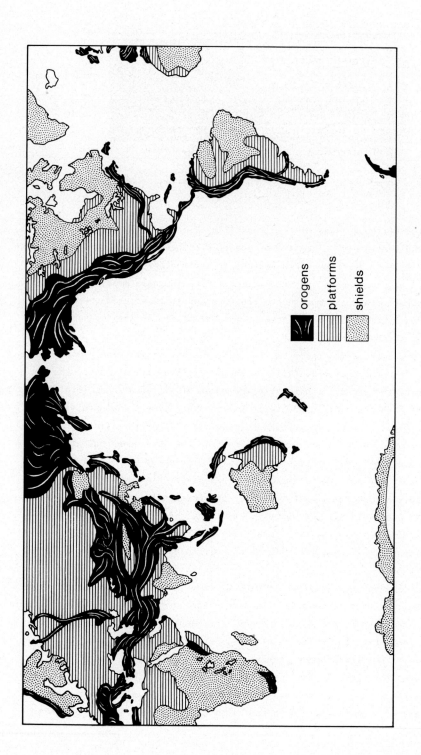

Fig. 2.2. Distribution of shields, orogens and platforms.

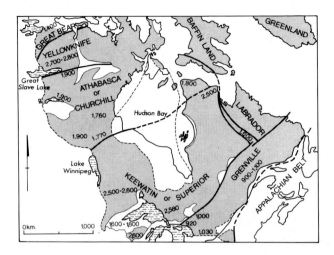

Fig. 2.3 (a). Structural units within the Laurentian Shield (after Holmes, 1965).

magma at varied depths in the crust or upper mantle and were intruded into earlier formed rocks. Regardless of their origins and the details of their composition, masses of magmatic granite are deep-seated bodies and, for this reason, are called *plutons*, after Pluto, the god of the underworld, the god of the dead in Greek and Roman mythology.

Plutons are classified principally according to the nature of their relation-ship with the country rock and only secondarily on their shape and size (Badgley, 1965, pp. 314-322). The margins of many plutons cut across structures in the adjacent rock and are therefore described as discordant. A few, like the Cairnsmore of Fleet pluton in Scotland (Parslow, 1968), have essentially con-cordant junctions, but in most instances the contact is partly discordant, partly concordant (e.g. Elders, 1963; Berger and Pitcher, 1970), and the bodies are classified according to which of the two situations is dominant. The following are the principal types of pluton:

Batholiths (or bathyliths) are massive intrusive bodies that are oval or shield-shaped in plan. Their diameter is maintained or increased to shallow depths, though there is some suggestion that they cut out within the crust (Bott and Smithson, 1967; Hamilton and Myers, 1967), indicating an overall lenticular or globular form (Fig. 2.5). Indeed, some exposed batholiths and stocks have a bulbous shape, with inward dips characterising the contacts (Fig. 2.6), again suggestive of a lenticularity. They may form a single extensive outcrop or several isolated exposures separated by country rock but linked in the sub-surface (Fig. 2.7).

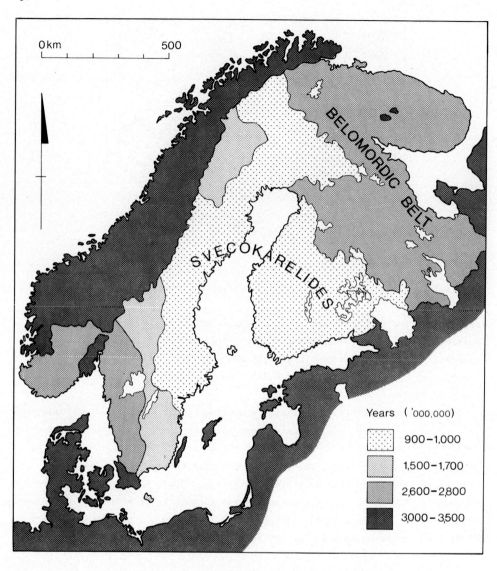

Fig. 2.3 (b). Structural units within the Baltic Shield (after Simonen and Mikkola, 1980).

Most batholiths are complexes that consist of many individual plutons. Thus the Sierra Nevada batholith of California, generated within a synclinorium and exposed intermittently over an area of some 60,000 km^2, in reality consists of as many as two hundred individual granitic bodies emplaced in phases over a period of 100 Ma (see e.g. Bateman et al., 1963; Bateman and Eaton, 1967 - also Fig. 2.5). The Blue Tier Batholith of northeastern Tasmania (Gee and Groves, 1971), the coastal batholith of central Peru (Cobbing and Pitcher, 1972) are

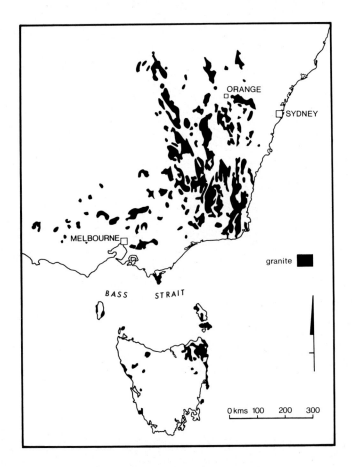

Fig. 2.4 (a). Occurrence of granitic batholiths in southeastern Australia (after Geological Society of Australia, 1971).

other examples of composite bodies.

Stocks are small batholiths (Fig. 2.8), conventionally less than 100 km^2 (40 square miles) in area. The contact between the granite and the host rock can be either steep (Fig. 2.9) or gentle.

Laccoliths are intrusive bodies that have caused doming of the roof rocks and are located at shallow depths, in relatively undisturbed areas (Figs. 2.8 and 2.10).

Phacoliths are sheet-like intrusive bodies situated in the crests of anticlines, where their upper and lower margins are convex upward (Fig. 2.11); they were intruded at the same time as the folding development.

Lopoliths are large lenticular intrusive masses, the central parts of which are sunken; the thickness of the bodies is between 5 and 10 percent of their width or diameter (Fig. 2.12).

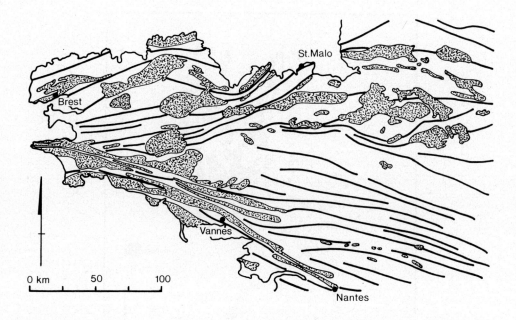

Fig. 2.4 (b). Occurrence of granitic batholiths in the northern part of the Armorican Massif, Brittany (after Bur. Rech. Géol. Min., 1968).

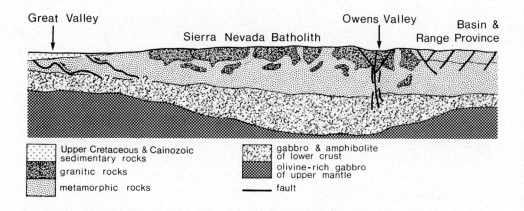

Fig. 2.5. Globular or lenticular form of various components of the Sierra Nevada batholith along the 37th Parallel (after Hamilton and Myers, 1967).

Gneiss domes are structural domes in granitic rocks (Fig. 2.13) and are probably developed by repeated uplift and intrusion.

Dykes (dikes) are discordant tabular bodies of intrusive origin, and cut across bedding or other existing penetrative structures in the country rock.

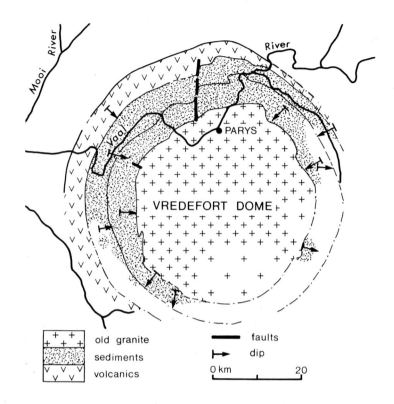

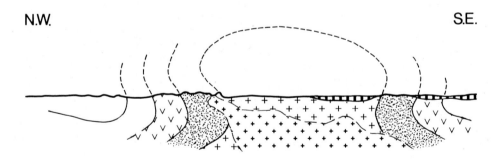

Fig. 2.6. Plan and section of Vredefort dome, Transvaal, R.S.A., showing centripetal dips at the margins of the granite body. (In part after du Toit, 1939).

Sills are tabular bodies emplaced parallel to the bedding, cleavage or foliation in the country rock.

Ring complexes are oval, circular or arcuate sills and dykes related to an intrusive centre (Fig. 2.14).

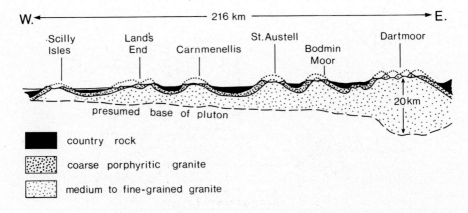

W. ◄──────── 216 km ────────► E.

·Scilly   Land's   St.Austell   Dartmoor
Isles   End   Carnmenellis   Bodmin
                                                   Moor

presumed base of pluton   20km

■ country rock

▒ coarse porphyritic granite

░ medium to fine-grained granite

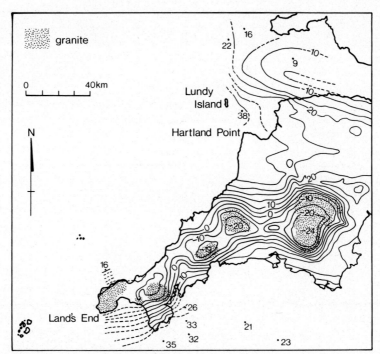

Fig. 2.7. Linked granite plutons of southwestern England above (a) in section (after Booth, 1968),   (b) below, indicated by Bouguer anomalies (after Bott and Smithson, 1967).

All of the major bodies described are of deep-seated origin. For though they may not now extend to profound depths, with increasing evidence that some batholiths for instance are of globular or lenticular shape rather than having their roots deep in the crust, all originated at considerable depths beneath the then land surface and all are plutons. They vary in plan shape and in areal extent,

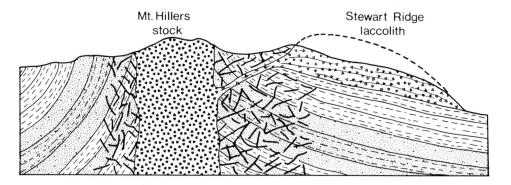

Fig. 2.8. Section through Mt Hillers stock and Stewart Ridge laccolith, Henry Mts, Utah, U.S.A. (after Hunt, 1953).

but they are in varying degrees discordant in relation to the structures of the surrounding rocks and all are composed of crystalline rocks with rocks of granitic composition most common.

Fig. 2.9. Plunging contact (A-B) of granodiorite stock (left) cutting Cretaceous sediments (right), Castle Peak, British Columbia, Canada (after Daly, 1912).

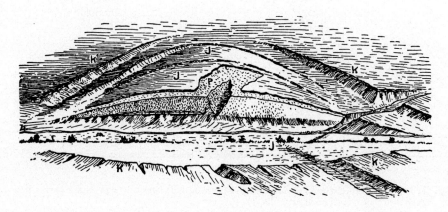

Fig. 2.10. Section through partly exposed laccolith, 1.5 km in diameter, Judith Mts, Montana, U.S.A. (after Weed and Pirsson, 1898). P - laccolith, K - Cretaceous, J - Jurassic.

B. DEFINITION AND COMPOSITION

   The word granite is derived from the Italian word *granulo*, meaning a grain or particle, and was first used by one Caesalpus in 1596. From the Renaissance onwards it was applied to all obviously crystalline rocks. By the middle of the Eighteenth Century, however, the word had a more restricted, technical, meaning so that Playfair (1802, p. 82), for example, could write of the greatest of the early, and one of the greatest of all, geologists:

   The term granite is used by Dr. Hutton to signify an aggregate stone in which quartz, feldspar and mica are found distinct from one another, and not disposed in layers.

Subsequent definitions have become more circumscribed and precise, and Hutton's preclusion of apparent stratification has, perforce, had to be disregarded in some areas (see below, Chapter 2 C ii (c)).

   According to Chayes (1957, p. 58) 'The name granite could usefully be reserved for massive or weakly oriented rocks' that are leucocratic and contain 'no less than 20, and not more than 40 per cent of quartz by volume'. Granites are plutonic macrocrystalline rocks that consist of interlocking crystals with few voids. Granites vary in composition, but true granites have at least 10% and up to 40% free quartz, together with feldspar and mica (see Streckeisen, 1967, 1974, for discussion of definition and classification of granitic rocks). Feldspars are the dominant rock-forming minerals and provide the basis for classification of granites and many other crystalline rocks. The feldspars are aluminosilicate minerals and form a continuous series between the potassic ($KAlSi_3O_8$), sodic ($Na.AlSi_3O_8$) and calcic ($CaAl_2Si_2O_8$) feldspars. The potassium and sodium rich varieties, together, form the alkali feldspars of which ortho-clase, microcline and albite are common examples. The sodium and calcium members

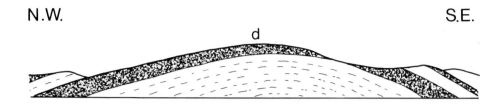

N.W.   S.E.

d

Fig. 2.11. Phacolith of dolerite (d) in anticline of Ordovician shale, Corndon, Shropshire, England (after Lapworth and Watts, 1894).

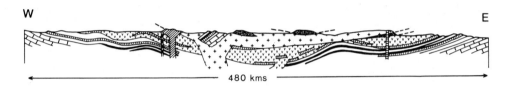

W   E

480 kms

Fig. 2.12. Section through Bushfeld Lopolith, Transvaal, R.S.A. (after du Toit, 1939). The structure is almost 500 km in diameter.

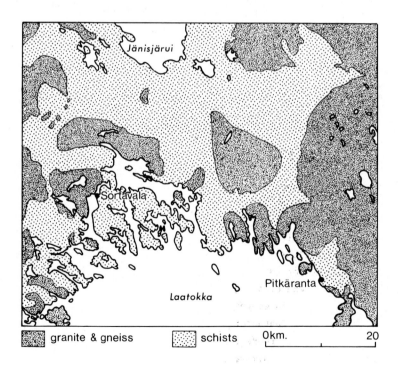

Jänisjärui

Sortavala

Pitkäranta

Laatokka

granite & gneiss   schists   0km.   20

Fig. 2.13. Gneiss domes near Lake Ladoga, southern Finland (after Eskola, 1949).

38

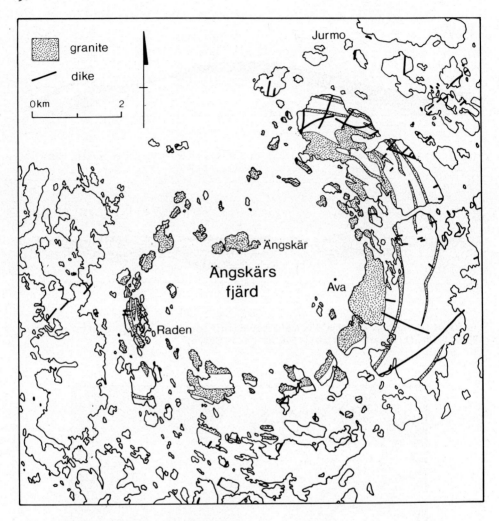

Fig. 2.14. Ava ring intrusion, Åland Islands, southwestern Finland (after
Kaitara in Simonen and Mikkola, 1980).

form the plagioclase feldspars (albite, oligoclase, andesine, etc.). Both alkali
feldspars and plagioclases are present in most granites. The relative frequen-
cies of alkali feldspar to plagioclase, together with grain size, form the
basis of subdivision and classification of granite and related rocks (Fig. 2.15).
Plutons are not uniform in composition. As has been mentioned, many igneous
masses are of composite character and there are differences on that account
(Fig. 2.16). In addition there are variations within individual bodies, such as
that reported from the Cairnsmore of Fleet pluton, where biotite content
decreases toward the centre of the mass (Parslow, 1968).

Many of the names given to granitic and associated rocks are self-explanatory. Others are of local derivation and are still occasionally used. Thus another name for granodiorite is tonalite after the Tonale Alps in northern Italy. The nearby Adamello Alps have given their name to adamellite. The syenites take their name from Syene (Aswan), in Egypt, and monzonite, a syenite that is well-known from its occurrence in the Yosemite Valley of California, takes its name from exposures near Monzoni, in the Tyrol. Unfortunately, some of these local names are no longer consistent with modern definitions: some rocks of the Adamello massif, for instance, are not considered to be adamellites but, rather, granodiorites.

In some granites all the crystals are of roughly the same size and their texture is described as equigranular (Fig. 2.17 (a)). Others are porphyritic (Fig. 2.17 (b)), that is, they display a strongly bimodal grain size, with some coarse crystals, typically of feldspar, set in a finer matrix or groundmass. Pegmatites are coarse-grained, in contrast with aplites which are fine-grained, granitic rocks, and both types usually occur as veins or dykes. Crystal orientation in granites is not random, feldspars and micas in particular commonly being aligned, probably reflecting primary flow structures; but lineation is not nearly so pronounced as in gneissic rocks, in which there is a secondary but distinct preferred orientation, or lineation (Fig. 2.17 (c)), and, in addition, a tendency for the micas especially to form distinct layers or folia that give distinct planes of preferred splitting that are known as foliation.

Granodiorite is by far the most common of the granitic rocks, being equal in areal extent to all other types together. The granitic rocks grade into a group of crystalline rocks in which free quartz is an accessory and not an essential mineral, which are again classified according to the ratios of the various contained feldspars, and which also display similar physical characteristics and landforms.

C. SOME PHYSICAL CHARACTERISTICS
(i) General

Granite is a crystalline rock with an average specific gravity of 2.662. A cubic metre of granite weighs of the order of 2,658 kg, or almost two tonnes a cubic yard. Its physical hardness varies according to composition, and principally with the proportion and type of feldspar present. Despite its crystallinity, granite is flexible in thin sheets. It also has considerable compressive strength.

Granite is characteristically of low porosity and permeability. In outcrops and near-surface zones, however, it is commonly fissured and fractured and is therefore pervious.

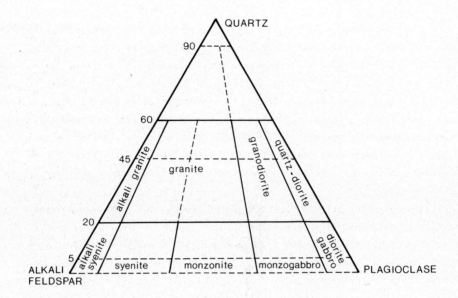

Fig. 2.15. Composition and classification of some common granitic rocks (after Streckeisen, 1967).

Porosity, also known in North America as mass permeability, refers to the ratio of the volume of voids to the total volume of the rock expressed as a percentage. Porosity varies with the shape of the constituent grains, as well as their sorting, packing and degree of cementation. A mass of closely packed uniform spheres consists of 26 percent by volume of pore space, but in a crystal-line medium such as fresh granite, the value is most commonly less than one percent. So far as granitic rocks are concerned, values vary not only because of the inherent character of different rocks but also because of differences in

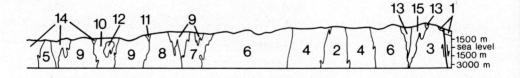

Fig. 2.16. Section through Sierra Nevada batholith west of Mono Lake, showing petrological variation within the composite pluton (after Bateman et al., 1963). 1 - sedimentary and volcanic rocks, 2 - granite porphyry, 3 - fine-grained quartz monzonite, 4 - granite (Cathedral Peak), 5 - miscellaneous granite rocks, 6 - quartz monzonite, 7 - granodiorite, 8 - granite (Taft), 9 - granite (El Capitan), 10 - granodiorite (Gateway), 11 - mafic plutonic, 12 - biotite granite, 13 - metavolcanics, 14 - metasediments, 15 - sediments.

Fig. 2.17 (a). Thin section of medium grained equigranular granite (scale in
mm) (A.R. Milnes).

the degree of weathering (see e.g. Dale, 1923; Kessler *et al.*, 1940). But
porosity is consistently low in granite, values of the order of 0.1 - 1.2 percent
being characteristic (Dale, 1923). Nevertheless, even fresh granite contains
some voids and a cubic metre of granite in the groundwater zone contains 15 -
20 litres of water in its pores.

The authors referred to tested only commercially-quarried granites from North
America; no doubt a greater range of values would be found were a global sample
available, but the orders of magnitude quoted are reasonable and porosity can be
taken as being typically very low in fresh, unweathered granites.

Permeability, also known in North America as primary permeability, refers to
the capacity of a porous medium to transmit fluids. It is not the same measure
as porosity, for voids may be unconnected, or too small to permit free passage
of liquids because of surface tension effects. Being crystalline, granites have
low transmission rates when fresh, though weathered rocks are commonly much more
permeable.

Perviousness is known in North America as secondary or acquired permeability.
The term refers to the capacity of a rock to transmit fluids, not through the
body of the rock, but by way of fractures. This property varies not only with
the number of fractures per unit of area (the density of the fracture pattern)
but also with their width - whether the cracks are tight, hairline fissures,

42

Fig. 2.17 (b). Porphyritic granite, South East district of South Australia.

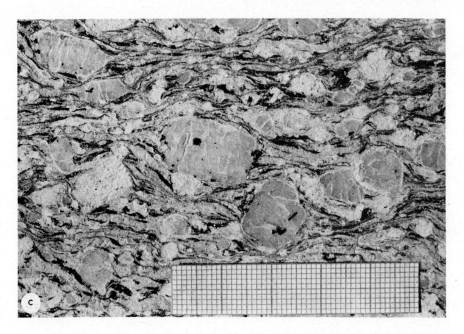

Fig. 2.17 (c). Thin section of gneissic granite with augen of potash feldspar and plagioclase set in layered matrix of biotite and quartz (A.R. Milnes).

or gaping.

The low porosity and permeability of granite near the land surface emphasises the importance of fractures and fissures, for they largely determine how readily water can infiltrate into the mass, there to bring about the alteration of the rock. The basic patterns of weathering and of landform development are thus in large measure determined by fractures.

(ii) Fractures

Like other crystalline rocks, granite is characteristically fractured. Fractures along which there is no demonstrable displacement are called joints, while those along which there has been differential movement, or dislocation, are called faults. But it is in many instances difficult to differentiate positively between joints and faults in granitic rocks and to identify faults in granite. As Hills (1963, p. 365) points out 'many fractures in granite are termed "joints" largely because evidence for fault displacement in them is not obtainable'.

Fracture patterns observed in the field are frequently complex, but two sets, one orthogonal, the other essentially arcuate, have been recognised in granites. The two seemingly overlap in part and they commonly coexist. In addition minor directions of easy splitting have been recognised.

(a) Orthogonal sets: Though they have been described earlier (e.g. MacCulloch, 1814; de la Beche, 1839, p. 450; Shaler, 1887-8), Hans Cloos (1922, 1923, 1936; see also Balk, 1937) was the first to see fractures in granitic rocks as integrated patterns. He named and classified fractures according to their geometric relationship with flow lines, platy flow structure

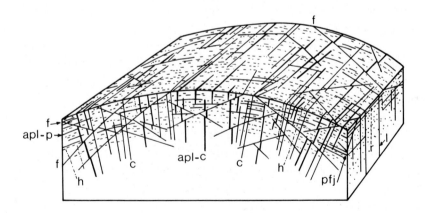

Fig. 2.18. H. Cloos' diagram showing the major types of fracture occurring in a batholith: c, cross joints; l, longitudinal joints; f, flat-lying faults some of which are planes of stretching; apl-c and apl-p, aplitic dikes; dashes, linear flow structure; r, rift; h, hardway (after Balk, 1937).

Fig. 2.19. Orthogonal joint sets exposed in quarry at Mt Bundey, near Darwin N.T.

or *schlieren*, and with foliation (Fig. 2.18). Thus, cross- or Q-joints (the
term joint is used without prejudice for fractures along which it is not
possible to demonstrate dislocation) lie perpendicular to flow lines, whereas
longitudinal or S-joints are parallel to them. *Lägerklufte*, or flat-lying
joints, are disposed normal to both Q- and S-joints. The three together effectively subdivide the mass into cubic or quadrangular blocks, depending on their
spacing. Together they form the well-known *orthogonal sets* plotted, for
example, by Shaler (1887-8) at Cape Ann, Massachusetts (see e.g. Figs. 1.6 and
2.19). They are also well developed in other types of crystalline rocks
(Fig. 2.20), and occur at various scales (Figs. 2.19 and 2.21).

Orthogonal fracture sets in granite are clearly related, in a geometrical
sense, to regional structural lines. Thus on northwestern Eyre Peninsula and
the adjacent Gawler Ranges the major NW-SE and NE-SW fractures that are prominent within, and that in some instances delineate, the major rock masses run
parallel to regional lineaments and fractures such as the Corrobinnie and
Kappakoola fault zones (Bourne *et al.*, 1974) and an offshore fracture zone that

Fig. 2.20. Orthogonal joint sets in gabbro, Giles Complex, northwestern S.A. (R.W. Nesbitt).

has been detected by seismic means (Sutton and White, 1968; Twidale, 1968a, p.54).

Again, in southwestern England the discrete plutons display fracture patterns that are clearly continuations of those in the surrounding sedimentary and metamorphic rocks. Such is the perfection of the regional patterns that they have the appearance of being overprinted on plutons and country rock alike. In French Guyana the pattern of fractures (Fig. 2.22) displays clear parallelism with regional lineaments such as that followed by the coast and by the Oyapok River (Choubert, 1974).

The orthogonal sets are thus conformable with regional fracture patterns and they also bear a positive geometric relationship with other structural axes. For this reason many writers consider that the fracture patterns have their origin in crustal stresses developed either during the emplacement of the crystalline masses, or imposed subsequently (see e.g. H. Cloos, 1936; E. Cloos, 1936). Workers such as Richter and Kamanine (1956) relate the major orthogonal sets of such areas as the Canadian and Baltic shields to regional concentric and radial fracture patterns, and writers like Carey (1955) to major shears in the crust. But, whatever the ultimate causes, the fractures are associated with shear couples.

46

Fig. 2.21. Large-scale joint sets forming a rhomboidal pattern in plan, and with gneissic foliation striking across the parallelograms, Hamilton district of central Labrador (Dept. Energy, Mines, Resources, Canada).

    (b) Rift, grain and hardway: In addition to planar fractures, it has long been recognised by practising quarrymen that granite (unlike most other crystalline rocks) splits more easily in some directions than in others. In the vertical or near-vertical plane three directions are recognised. Rift is the direction of easiest splitting, and is also known as reed and cleaving way. Grain is the direction of fairly easy splitting, though it is not as easy as rift; it is also known as hem, quartering way and cut off. Hardway is, as its name suggests, the direction of most difficult splitting.

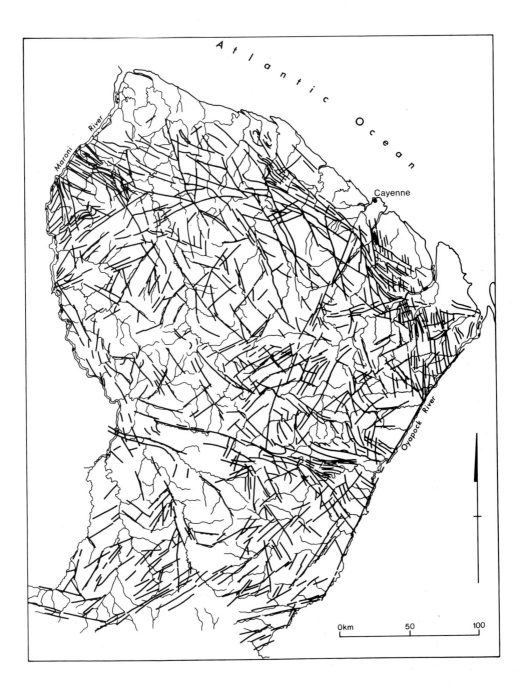

Fig. 2.22. Fracture patterns in French Guyana (after Choubert, 1974). Only major fractures and drainage lines are shown.

Some workers have given the name 'lift' to horizontal planes of easy splitting. These clearly run parallel to flaggy or sheeting structure (see below) and, according to several authors, the rift is frequently, though not everywhere, disposed parallel to vertical members of the orthogonal set. Cloos (1922), for example, suggests that rift and S-joints are usually parallel. Tarr (1891, p. 270), though less specific, states that in a number of cases 'the rift is found to follow the same general direction as the principal set of joints'. It is fair to add, however, that rift varies in direction even in the same quarry (though it runs parallel to vertical joints), and that some authors disclaim any relationship between rift and joint directions (e.g. Osborn, 1935).

According to Tarr (1891) and Dale (1923, pp. 15-26), rift structures consist of minute fractures which, under the microscope, are seen to be faults since small dislocations are detectable (Fig. 2.23). In places, the fractures deviate to follow platy cleavage and foliation, as well as other inherent weaknesses such as crystal cleavage. Rift also runs parallel to bands of cavities in quartz grains, but, viewed as a whole, the fracture planes, each comprising 'thousands of minute dislocations which occur in every cubic inch of rock' (Tarr, 1891, p. 269), cut across rock texture.

As to origin, Tarr (1891) considers that rifting and jointing are genetically related, that both are a manifestation of rock stress, and that both are due to 'faulting and contortion' (Tarr, 1891, p. 268). Rift is thus comparable with crystal dislocations, due to stress, noted in olivine for instance (Boland *et al*., 1971). Rift could have developed before, after, or contemporary with the joint fractures.

Thus both regional relationships and microstructures strongly suggest, first, that the orthogonal joints in granite are related to rock stresses and, second, that most of the fractures are very small displacement faults.

(c) 'Bedded' granite: Various types of fractures running parallel to the land surface have been described in the scientific literature for almost three centuries (Dale, 1923, p. 26). Some joints which parallel the land surface are closely spaced, extend to only shallow depths, and are associated with minor facets of the land surface that are, in a geological context, only youthful. These are the fractures variously referred to as exfoliation (Matthes, 1930; Harland, 1957), pseudobedding (e.g. Waters, 1954) and lamination (Twidale, 1971a). None of these terms is satisfactory. Worth (1953) used the term 'lamellar' which, though reasonable when applied to thin flakes or scales, is inadequate to describe the thicker (10-20 cm minimum) slabs that are really flagstones in granite (Fig. 2.24).

Such bedded or flaggy granite differs in its distribution, both in space and time, from the thicker, deeper sheet structure. It is, for example, frequently

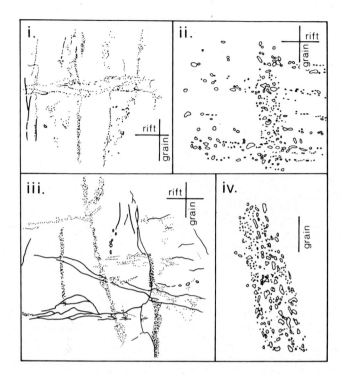

Fig. 2.23. Microscopic structure of rift and grain (after Dale, 1923).

discordant with orthogonal jointing (Fig. 2.25). Flaggy granite is well-known from the various granite outcrops in southwestern England (see e.g. Waters, 1954); from the Karkonosze Mountains of southern Poland (Jahn, 1974); from Mongolia (Dzulinski and Kotarba, 1979); from New England (U.S.A.), whence Shaler (1887-8) described what he called the 'extreme development of horizontal jointing' in the Pigeon Hill Quarry in Massachusetts; and from many other cold and cool regions. It is essentially a superficial feature, though it extends to depths of several metres in such areas as Dartmoor and Spitzbergen.

On the other hand, the term flaggy is scarcely appropriate for those thin layers that seemingly wrap around domical hills, as in the Yosemite, and for these the term spall plate may usefully be employed.

(d) Sheeting joints: In addition to these near-surface flaggy structures, Dale (1923), Ljungner (1930) and others recognised that sets of horizontal or curvilinear joints trending approximately parallel to the land surface extend to depth in many localities. They have been given different names: flat-lying joints, *Lägerklufte, Bankung, structure en gros bancs*, stretching planes, shells, exfoliation, sheeting joints or sheet jointing, and offloading, relief of load, or pressure release joints. Here, the term *sheeting joint* (and its variant

Fig. 2.24. Flaggy structure (a) in granite on the summit of Roughtor, Dartmoor (Geol. Surv. Mus. U.K.), (b) in monzonite, Sierra Nevada, California.

sheet jointing) is used for no other reason than that it can be used in a purely descriptive sense, with no genetic implications or prejudices derived from earlier interpretations. *Sheet structure* is used of the massive slabs defined by sheeting joints (Fig. 2.26).

Sheet structure and jointing have been described in some detail elsewhere (see, for instance, Dale, 1923; Jahns, 1943; Twidale, 1964, 1971a, pp. 59-90,

Fig. 2.25. Orthogonal joint sets in granite at Heltor, Dartmoor, southwestern England, with superficial flaggy structure or pseudobedding clearly discordant with it.

1973), so that there is no necessity to go into detail here. It is typically developed in granite, but is well-formed also in gneiss, dacite and rhyolite, sandstone, arkose, conglomerate and limestone (see e.g. Bradley, 1963). It occurs in both cratonic and orogenic regions as well as in strata that have suffered only minor flexuring.

Sheet structure is up to 10 m thick and is characteristically developed on massive rocks, that is, rock masses which, though not lacking fractures, do not display open joints. It has been described from every conventionally defined climatic regime.

Sheet jointing has been observed to depths of at least 100 m in several quarries, as, for instance, at Quincy, Massachusetts (Dale, 1923, p. 23), Barre (Vermont) and Merrivale, on Dartmoor, in southwestern England (Twidale, 1971a, p. 63), and is well developed in many other sites. The thickness of the sheets generally increases with depth, though there are many exceptions. According to Richey (1964), it fades in depth in some quarries. In the Yosemite Valley of California, and in New England, Australia (Leigh, 1967), sheet struc-ture is displayed in the rock masses exposed in the side slopes of deeply incised river valleys, suggesting that the fractures may extend to depths of some hundreds of metres below the crestal or summit surface; but whether the joints extend any distance into the hills or whether they are developed only near the valley sides is not known.

Fig. 2.26. Ucontitchie Hill, northwestern Eyre Peninsula comprises several massive arcuate slabs.

Sheeting joints commonly run parallel to the land surface and in valley floors or on broad crests are either planate or only gently arcuate. But on valley sides and in the vicinity of steeply-dipping faults, sheeting joints are also steeply dipping (Fig. 2.27). Vogt (1875) described sheeting which, though sensibly horizontal on hill crests, dips at angles of up to 33° toward valleys. The Devil's Slide on Lundy Island, in the Bristol Channel, is such a steeply inclined sheeting plane. On Ucontitchie Hill, a bornhardt on northwestern Eyre Peninsula, South Australia, the sheeting joints are almost flat on the crest (Fig. 2.28) but dip at angles of 70° near a major near-vertical fracture zone that delimits the residual on its eastern side (Figs. 2.28 and 2.29). Pronounced steepening of sheet structure in relation to major vertical joints has also been observed on Pearson Island (Fig. 1.22); and on the granite domes in many other parts of the world.

Some so-called sheets are in reality wedges, though whether this reflects the original fracture patterns or subsequent erosion is not known. In some places it can either be observed or inferred that the sheeting joints dip into hills at inclinations greater than the angle of slope (Figs. 2.27 and 2.30), suggesting that the hill crests, at any rate, are underlain by synformal wedge-shaped or lenticular masses of rock, as was indeed appreciated late in the last century by Herrmann (1895). Often the slabs of rock are separated from the parent mass by gaps several centimetres deep.

Wedges of rock up to one metre across and triangular in cross-section occur at the exposed outer edges of many sheeting joints (Fig. 2.31). Some of the wedges remain attached to the parent mass, but others are detached and some have been moved sideways. In many areas, for instance, southeastern Brazil (Lamego, 1938), Vermont (Fig. 2.32) and the southwestern parts of the Gawler Ranges, South Australia, sheet structure and faulting occur in association.

Fig. 2.27. Many sheeting planes and structures are exposed in the cirque head-wall in the Little Shuteye Pass, Sierra Nevada, California (N.K. Hubert). Note the synformal structure near the crest.

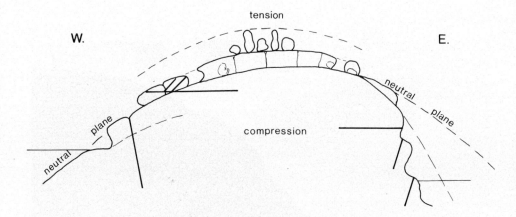

Fig. 2.28.   East-west section through Ucontitchie Hill, northwestern Eyre
Peninsula, S.A., showing steeply plunging sheet structure and flared margins.

Fig. 2.29.   Multiple flares developed on eastern slopes of Ucontitchie Hill and
cutting across steeply dipping sheeting joints.

Sheeting joints cut across other structures in the bedrock, such as orthogonal joints (Fig. 2.24), cleavage, rift and grain, and flow structures (see, for instance, Cameron, 1945; Richey, 1964; Twidale, 1964; Soen, 1965). They thus postdate the consolidation of the rocks, but in some areas they demonstrably predate subsequent phases of igneous injection, for on Dartmoor, for instance, the sheeting planes are followed by veins and sills which were intruded into the existing granite mass during the Mesozoic (Darnley, in discussion of Blyth, 1962). Elsewhere, however, as in Norway and the western United States, sheet jointing is of recent geological age (Kieslinger, 1960), though in New England (U.S.A.), Scotland and Ireland (Jahns, 1943; Richey, 1964) the sheeting predates the most recent local glaciation, for the sheet structure has been eroded by ice.

Fig. 2.30. This low dome near Tenaya Lake, Yosemite Valley, California, is apparently underlain by arcuate fractures (X) of synformal geometry.

Sheet structure is well and widely developed in granitic and indeed in many other massive rocks. It is commonly found to coexist with orthogonal sets, one being dominant at some sites, the other elsewhere. The origin of sheeting joints is discussed in a later chapter (Chapter 5) in connection with the evolution of bornhardts. Suffice it to say here that the association of the inselbergs and arcuate fractures may not be coincidental.

In summary, granite is an acid crystalline rock that is widely distributed in shields and orogens, in which contexts it is found in bodies of various shapes and sizes most of which, however, can be referred to as plutons. When fresh, granite is of low porosity and permeability but is usually pervious by

Fig. 2.31. Wedges (a) above, *in situ* on the western side of Ucontitchie Hill, northwestern Eyre Peninsula, S.A., (b) below, displaced, on the eastern flank.

virtue of its fracture systems.

Granite consists of quartz, feldspars and mica. The last two of these in particular readily react with the water that penetrates along partings, altering and also rendering permeable, the rock with which it comes into contact. Water then infiltrates deeper into the erstwhile cohesive rock which is rapidly

Fig. 2.32. Part of the Rock of Ages Quarry, Barre, Vermont, U.S.A., seen in early 1966, showing sheet structure and fault planes (f) cutting across the sheeting fractures.

reduced to a mass of altered material. Thus because of the susceptibility of granite to moisture attack the patterns of fractures greatly influence the cause of weathering or erosion, and hence of landform development.

The changes brought about by water in contact with rocks are some of a group of processes together referred to as 'weathering', the nature of which is discussed in the following chapter.

CHAPTER 3

WEATHERING WITH PARTICULAR REFERENCE TO GRANITIC ROCKS

Weathering can briefly be defined as the alteration and/or disintegration of
rocks *in situ* and in the range of ambient temperatures found at and near the
earth's surface. Some weathering processes are mechanical or physical, and
cause the disaggregation of the rock without the decomposition of any of the
constituent minerals. Chemical processes, on the other hand, involve the
alteration of one or more of the rock-forming minerals. Chemical and physical
processes are usually active at the same site and work together to bring about
rock disintegration, though various forms of chemical attack associated with
water are the most potent and widespread in the weathering of granite.

The destructive aspects of weathering are emphasised, and rightly so, for
weathering is an essential precursor to erosion. But chemical weathering
releases into solution salts that remain within weathering profiles and within
drainage basins for extended periods. Some are eventually precipitated to form
distinctive concentrates of considerable geomorphological and economic significance. Thus, weathering has a positive as well as a negative aspect. Case
hardening (see Chapter 10) provides a minor but important example of the positive
or constructional role of weathering. Even small concentrations of the oxides
of silicon, and in places of iron and aluminium, are enough to endow altered
granite with a sufficient resistance to form cappings and give rise to plateau
forms (Fig. 3.1). In addition, some granitic rocks have been weathered enough

Fig. 3.1. This mesa stands in an anticlinal valley in the core of which granite
is exposed in the well-known Devil's Marbles, N.T. The mesa is eroded in
granite and is capped by granite in which the feldspars are altered to kaolinite
but with goethite(?) filling the intergranular spaces. In the background are
bevelled sandstone ridges, remnants of an Early Cretaceous planation surface of
which the mesa is part.

for laterites, bauxites and silcretes to have evolved on them. The commercial
bauxites of the Darling Ranges, Western Australia, for instance, are largely
developed on granite. After dissection, such duricrusts become desiccated,
form caprocks (Fig. 3.2) and substantially protect and preserve what are in
some instances very old land surfaces (Twidale, 1976a).

Fig. 3.2. Plateau, mesa and butte forms capped by laterite developed on
granitic rocks, southern Isa Highlands, northwest Queensland (Div. Land Res.,
C.S.I.R.O.).

The lower or lateral limit of significant weathering is called the *weathering
front* (Mabbutt, 1961a). It separates the still fresh rock from the weathered
mantle, saprolith or regolith, though the latter is usually taken to include
the veneer of transported alluvial and colluvial debris that might be present.
The weathering front may be essentially planate and simulate the form of the
land surface, but in detail it is irregular with protruberances of fresh rock
here and minor depressions there (see Chapters 5, 8 and 9). In granitic rocks,
which are usually of low permeability, there tends to be a sharp break, at most
a narrow zone of transition, between weathered material and essentially fresh
rocks (Fig. 3.3). Many granite forms are essentially exposed weathering fronts
or *etch* forms (Wayland, 1934; but see also Jutson, 1914).

Granite occurs in the same range of environments and is subject to the same
range of weathering processes as are other rock types. It is, however, particularly vulnerable to moisture attack, but is relatively stable when dry.

Fig. 3.3. The weathering front in these granitic rocks is marked by an accumulation of iron oxides leached from the one above.

Darwin (1846, p. 428) appreciated this contrast in southeastern Brazil, for he wrote that 'The porphyritic gneiss, where now exposed to the air, seems to withstand decomposition remarkably well'. De Saussure (1796, I, p. 518) had earlier hinted at similar behaviour and the notion, involving positive feedback or reinforcement effects, has been taken up by several writers during this century (e.g. Barton, 1916; Bain, 1923; Twidale, 1962; Wahrhaftig, 1965; Twidale *et al.*, 1974).

On a regional canvas the implication of this susceptibility to moisture attack is that in humid climates, past or present - and there are no areas that either have not received or do not receive substantial rainfall or accessions of groundwater - granite tends to be altered, so that fresh granite can be considered the exception rather than the rule. It is masses of fresh rock, not zones of weathered grus, that require explanation.

In this chapter various weathering processes affecting granitic rocks are briefly and generally discussed, as are some of the landforms resulting from their operation. Further reference is made to weathering processes in relation to specific landforms at appropriate places in later chapters. Here, general principles are discussed.

A. PHYSICAL WEATHERING

Earlier generations of geomorphologists set great store by the physical disintegration of granitic and other rocks. In particular it was argued that heating would cause wedging and straining as minerals with different coefficients of thermal expansion were heated (e.g. Tarr, 1915; Sosman, 1916). It is true that the ephemeral, but intense, heat generated in bush (or forest) fires causes superficial flaking (Fig. 3.4). In places the flakes are several centimetres thick, as, for instance, at one site observed by the writer in the Reynolds Range, north of Alice Springs, N.T., and near Tampin, in West

Fig. 3.4. Large residual granite boulder with superficial fire flaking,
Mt Manypeaks, near Albany, southwest of W.A.

Malaysia (Fig. 3.5), but more commonly they are less than a centimetre thick
and have distinctive paper-thin crenulated edges (see also Blackwelder, 1927;
Emery, 1944). The intense heat of atomic explosions has similar effects
(Watanabe *et al.*, 1954), and the heating of granite slabs was a method of
quarrying practised in ancient India and Egypt (Warth, 1895).

But disintegrated granite extends to depths far beyond the reach of ephemeral
intense heating, or of diurnal or of annual temperature changes, and this consideration, plus the observations, experiments and arguments of such workers as
Blackwelder (1925, 1933), Griggs (1936) and especially Barton (1916) very
strongly suggest that insolation alone is, at best, a minor and contributory
factor in rock disintegration.

Temperature oscillations in the presence of moisture are another matter.
Griggs (1936) suggested as much, Birot (1950) thought that heating and immersion facilitated intercrystalline aqueous circulation and in particular the
alteration of biotite, and Moss *et al.* (1981) achieved some fragmentation by
subjecting granite detritus to temperature cycling between $-4.5^{\circ}C \pm 1.0^{\circ}C$ and
$13.0^{\circ}C \pm 2.0^{\circ}C$ in the presence of added water. A similar disintegration was
achieved, however, in the presence of absorbed water.

Fig. 3.5. This large residual boulder near Tampin, West Malaysia, is pitted from base to crest, a distance of about 2 m, indicating that it was, until recently, immersed in the regolith and that it has recently been exposed. Thus soil erosion of about 2 m is indicated. Local information suggests that it has taken place during the last 10-15 years. The rain forest has been cleared to make way for rubber trees, and the felled timber burnt on the spot, causing lenticular plates to flake off the main mass exposing a smooth granite surface. Some of the spall plates can be seen resting against the base of the rock.

Alternations of freezing and thawing seem more promising as a mechanism causing disaggregation, especially where the granite is finely fissured and fractured. Thus, according to many writers (e.g. Orme, 1964) the clitter and angular blocks of Dartmoor and other cold regions (Figs. 3.6 and 3.7) is produced by the frost riving of granite already split into flaggy slabs and spall plates. Flaggy granite (Chapter 2) is undoubtedly best developed in areas that are, or recently have been, cold, and there is little doubt that Harland (1957) is correct in attributing its development largely to the freeze-thaw mechanism.

Fig. 3.6. Frost-riven plates of granite at an altitude of about 3,600 m in the Rocky Mountain National Park, Colorado.

Fig. 3.7. Frost-riven blocks in the Andorran Pyrenees. They have obviously been split along major fractures, and presumably by frost action. But these blocks are of considerable mass. Is freezing water capable of moving such large blocks? It may be that innumerable minor oscillations of temperature around the critical freezing point gradually accomplish this, but there is, nevertheless, a difficulty.

In road cuttings in the Adirondacks, for instance, there are many examples of
such flaggy partings, with water seeping from them in summer, but filled with
ice in winter. Spall plates follow around all but the most recent land surfaces, and, for this reason, are commonly considered to be of relatively recent
age.

Over the years, however, several writers have questioned the effectiveness
of the freeze-thaw mechanism. Many fissures and fractures are open to the air,
so that the pressures exerted by ice are unconfined. Many granite masses are
massive: in other words, there are few partings along which water could penetrate. Many of the blocks delineated by fractures are so large and heavy that
they are surely beyond the power of frost expansion to move or disrupt (Fig. 3.7).
Pressures in adjacent partings, particularly vertical or near-vertical fractures,
are opposed and therefore ineffective. Grawe (1936) noted that estimates of
the pressure exerted by frozen water and the nature of materials affected (con-
solidated or unconsolidated) vary widely. The freezing of water causes a
pressure change and the depression of freezing point. The change of state
liberates heat. Bridgman (1912) had earlier pointed out that at $-22^{o}C$ there
was a change in the crystallinity of the ice to a lower volume form, though this
may be irrelevant insofar as the intensity of freezing is not so important as
the frequency of temperature oscillations around $0^{o}C$ (see e.g. Potts, 1970).

Thus, though the mechanism is not rejected, there are several difficulties,
and it may be that it works best in conjunction with other processes.

B. CHEMICAL ATTACK

There is some suggestion that mineral reactions take place at grain contacts
in dry conditions, with illite and kaolinite formed between muscovite and orthoclase and muscovite and plagioclase respectively, and with quartz stable
(Meunier and Velde, 1976 - see also Fig. 3.8), but penetration by moisture
produces more rapid, marked and widespread effects.

Like most other rocks, granite is susceptible to attack by water, for all
minerals react with water to some extent. Because of its particular molecular
structure (Mason, 1966, p.161), water is the supreme solvent. No other liquid
can dissolve such a variety and volume of solutes. It will be recalled
(Chapter 2) that even fresh granite may contain considerable volumes of interstitial or pore water, but its effectiveness as a weathering agent depends on
whether it is static or circulating, its temperature, and so on. All minerals
are soluble in some measure, and for this reason it has been claimed that
'solution is essential to chemical weathering' (Loughnan, 1969, p. 61). In
addition to its direct effects, solution prepares crystal structures for
further reactions and, in particular, for hydration (combination of water

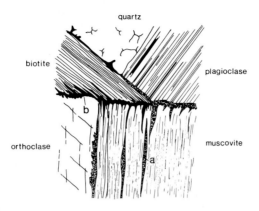

Fig. 3.8. Secondary mineral development through interaction at crystal boundaries, (a) micaceous phase, (b) diffusing zone (after Meunier and Velde, 1976).

with the mineral) and for hydrolysis (combination with OH radicles). In general terms, a granitic rock consisting of quartz, a potassium feldspar plus plagioclase and a mica, is altered to kaolinite plus an expanding mineral and an oxide or hydroxide, but there are many variations. Under continental conditions, two contrasted processes - leaching and fixation - are in competition (Nesbitt et al., 1980; also Nesbitt, 1979). During the degradation of clays, cations are leached from the primary minerals, but later some of these same cations are fixed by exchange and adsorption on to the secondary clay minerals. Ca, Sr and Na are most rapidly dissolved and, although large quantities of Mg are also evacuated, considerable amounts of elements like Rb, Cs and Ba remain in the weathering profile fixed to the secondary clays. Thus water is an important medium of transportation, as well as a solvent.

The solubility of any mineral varies both inherently and with the characteristics of the solvent. Most minerals increase in solubility with rise in water temperature, and quartz-rich rocks are more susceptible to attack by alkaline waters than they are to faintly acidic solutions (see Pouyllau and Seurin, 1982; also Joly, 1901; Mason, 1966, p. 142; Twidale, 1980a), only very highly alkaline (pH 9+) waters are aggressive however. This condition is found in some places but only very rarely in granitic environments, where the nature of the chemical reactions generally produces neutral or faintly acid environments (pH typically 4.0-6.0).

The reason for this is that water reacts with feldspars to produce clays (see below). In many instances the change involves hydrolysis. This, in turn, implies the disassociation of water and release of hydrogen ions. These combine high energy with a small ionic radius which makes them active in substitution. Hydrogen ions readily enter and disrupt crystal lattices (that is, they

cause alteration), but the presence of abundant hydrogen implies acid soils.
Moreover, and as is made clear in later chapters, much of the weathering considered here took place in the Mesozoic and earlier Cainozoic, when humid warm
conditions prevailed and when podsolisation, expressed in widespread lateritisation and ferruginisation, occurred. In such warm humid conditions water in
capillary pores becomes disassociated and faintly acidic. Plant roots assist
in the formation of acid clays by causing a concentration of hydrogen ions
(e.g. Keller and Frederickson, 1957). For all of these reasons, then, acid
weathering and soil conditions are implied in granitic terrains - acid conditions that are moreover conducive to silica solution (see Raupach, 1957).

Quartz is rightly regarded as the most stable of the common rock-forming
minerals and certainly of those that occur in granite, but silica is slightly
soluble in water and is a consistent and significant component of the dissolved
load of rivers the world over (Davis, 1963). Such solution attack is not confined to the crystal faces. Moss and his co-workers (Moss, 1973; Moss *et al.*,
1973; Moss and Green, 1975) have shown that quartz grains from granites are
characteristically riddled with cracks or microfractures (Fig. 3.9) that are

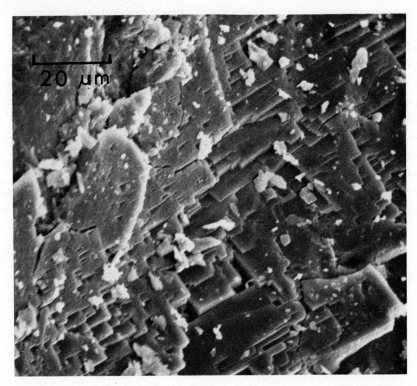

Fig. 3.9. Detail of orthogonal microfissures induced by a hammer blow and
developed in granitic rock from the A.C.T. (A.J. Moss, Division of Soils,
C.S.I.R.O.).

most likely due to stresses imposed during emplacement or during subsequent
tectonism. They claim that the grains split during transport and attribute
this disintegration to mechanical attrition; but it may be that, in addition,
water penetrates the cracks and either by solution or by hydration (see below)
causes the bonds between adjacent fragments to be weakened (see also Moss et
al., 1981). Such water penetration has other effects. According to Eggleton
and Buseck (1980) solution takes place primarily at the junctions of contrasted
lattice structures, where there is high strain, and possibly dislocation due to
stress (see Boland et al., 1971). Thus the borders of microcline included in
albite are the loci of initial solution in one of the rocks studied. Moreover,
the dislocations caused at the edges of lattices by the initial solution spread
and expand through the adjacent particles, thus permitting readier access to
water, and further alteration.

This is in keeping with the suggestion, due to Turner and Verhoogen (1960,
p. 476), that, all else being equal, stressed rocks are more readily dissolved
than are those that are relaxed. But, whatever the reason, the quartz of
granitic rocks is attacked and fragmented by groundwaters, and though it persists in weathering profiles longer than any of the other essential rock-forming
minerals it, too, eventually disintegrates.

Its reactions, however, are slight and slow compared to those of biotite and
feldspar. Biotite is particularly susceptible to moisture attack (Birot, 1950).
In a moist environment the mica takes water into its lattice and expands,
particularly along basal cleavages (Isherwood and Street, 1976). The fractures
developed in the expanded biotite extend as microfissures through adjacent
quartz and feldspar crystals. The biotite is altered to hydrobiotite, and
biotite/hydrobiotite interlayers (though other forms of mica, in different
conditions, are changed to vermiculite, chlorite, kaolinite, depending on
circumstances). The chemical changes are slight, involving oxidation of Fe^{+2}
to Fe^{+3} and the replacement of potassium by water molecules, but the physical
effects are notable and the implications, in the way of increased accessibility
to water, enormous. Indeed, Isherwood and Street attribute the disintegration
or 'grussification' of the granodiorite they studied to biotite alteration and
expansion.

Feldspars react with water and undergo argillation to give amorphous colloids,
oxides and clays, the precise nature of the latter depending broadly on climate.
Generally, in the humid tropics and humid temperate lands the end-product of
feldspar and biotite alteration is most commonly kaolinite, though gibbsite is
occasionally attained in tropical regions. Montmorillonite is commonly formed
in the arid topics, as it is in poorly drained sites in mid-latitude lands.
But, again, there are many variations according to local environmental conditions, particularly moisture regimes, past and present.

For example, in detail, drainage effects are significant. The hydrolysis of a potash feldspar may produce either illite or kaolinite according to whether the potash produced by the reaction is removed from the system. In conditions of good drainage, the potash may be evacuated:

potash feldspar + water   kaolinite + silica + potash

$$2\ KAlSi_3O_8\ + H_2O\ \longrightarrow\ Al_2Si_2O_5(OH)_4\ +\ 4SiO_2\ +\ KOH$$

If some potash remains in the system however:

potash feldspar + water   illite   + silica + potash

$$3KAlSi_3O_8\ + 2H_2O\ \longrightarrow\ KAl_2(AlSi_3)O_{10}(OH)_2\ + 6SiO_2\ + 2KOH$$

The typical end-product of the chemical weathering of granitic rocks is a gritty, puggy clay. The grittiness is due to the contained fragments and crystals of quartz. The clays, the products of the alteration of feldspars and micas, are coloured red or brown or grey, according to the degree of oxidation of iron.

C. INITIAL BREAKDOWN

Chemical weathering, and particularly various processes of alteration caused by reaction with water, causes the breakdown of granite. The rock is, however, characteristically impermeable. Moreover, several writers have shown (Tables 3.1 and 3.2) that the chemical differences between disaggregated and fresh cohesive rock are not great; apart from a slight concentration of silica in the grus, the other changes are minor. Yet it is assumed that these changes, though they are evidently slight, are sufficient to rupture the rock. 'A slight hydration of biotite and other minerals is probably sufficient to effect the change in volume that produces the disintegration and formation of boulders' (Larsen, 1948, p. 115). Anderson (1931, p. 59) attributed the weathering of granite in Idaho to 'solutions which penetrate slowly along the cleavage cracks in crystals and between mineral grains and there induce the formation of minerals of larger volume'. He thought that the disintegration of the granite was due to moisture attack, adding that, though 'the degree of decomposition appears to be slight', it was, nevertheless, sufficient to cause disaggregation.

Similar results have emerged from studies in South Australia. Examination of disaggregated granite from Palmer in the eastern Mount Lofty Ranges showed that, apart from very small amounts of clay developed along cleavage in the orthoclase, there was no detectable alteration of the rock-forming minerals: the quartz was fresh, the biotite unchanged, yet the rock was friable and weak. Very little alteration can be detected in flaked granite from Eyre Peninsula; studies of gabbro at Black Hill in the western Murray Basin in South Australia showed that the flakes and thin slivers of rock were coated with clay (Hutton et al., 1977).

TABLE 3.1   (after Larsen, 1948, p. 116)

Comparison of fresh and weathered granodiorite, Mt Woodson, California.

	Fresh boulder	Grus 3 m (10 feet) from surface	
SiO_2	74.72	76.13	+1.41
TiO_2	0.16	0.16	-
Al_2O_3	13.72	12.97	-0.75
Fe_2O_3	0.41	1.17	+0.76
FeO	1.32	0.35	-0.97
MnO	0.01	tr	-
MgO	0.23	0.07	-0.16
CaO	1.62	1.40	-0.22
Na_2O	3.76	3.79	+0.03
K_2O	3.40	2.70	-0.70
H_2O-	-	0.09	+0.09
H_2O+	0.33	0.78	+0.45
CO_2	-	-	-
P_2O_5	0.03	0.03	-
S	0.01	0.04	+0.03
BaO	-	0.09	+0.09

TABLE 3.2 (after Chapman and Greenfield, 1949, p. 424)

Comparison of relatively fresh and slightly weathered biotite granite from Virginia Dale, Colorado.

	Fresh	Weathered
perthite and microcline	36.5	41.0
oligoclase	23.1	25.2
quartz	31.1	24.6
biotite	6.2	4.7
sericite	1.6	3.3
magnetite	1.5	1.2
apatite	-	tr

Fig. 3.10. Corestone bordered by thin flakes intervening between fresh rock and grus, Palmer, S.A.

Despite the obvious efficacy of chemical, and particularly moisture, attack, there are grounds for believing that the initial breakdown of granitic rocks is, in part at least, mechanical. The fresh rock is frequently observed to br fringed by flakes, scales or spalls that represent the first stage in the weathering of the mass (Fig. 3.10). Larsen (1948, p. 115) stated that in southern California 'Between the boulder and the gruss there is commonly a thin layer made up of shells parallel to the surface of the boulder'. Similar transitions have been noted in the South Cameroon (Boyé and Fritsch, 1973) and many other examples have been noted in the field, as, for instance, at various sites on Eyre Peninsula; at Palmer and at Black Hill (norite) in the eastern Mount Lofty Ranges and adjacent parts of the Murray Basin respectively; the Rocklin Quarry in the western foothills of the Sierra Nevada, California; and Paarlberg Quarry near Cape Town. There are minute amounts of clay between the slivers at some sites, but whether these cause the fracturing or follow the penetration of water along the partings is not clear. Once fractured, however, water can readily penetrate and disintegration and alteration inevitably follow.

One possible explanation of this and other forms of disaggregation is the crystal dislocation resulting from solution demonstrated by Eggleton and Buseck (1980) and already discussed. Another is what has been called hydration shattering (the *Hydratationssprengung* of Wilhelmy, 1958, p. 52). According to White (1973) this does not involve the reaction of any of the minerals with water, but rather the addition of water molecules to the mineral structure.

In these terms hydration shattering occurs because:

> Surface boundaries of all silicates are electron clouds of oxygen atoms [which] attract water molecules. Polar charges orient the water molecules around them, adding a layer known as adsorbed or ordered water. This in turn attracts more water and builds up layers of adsorbed water of varying 'thicknesses' on the mineral surface. (White, 1973, p. 3)

Moss *et al.* (1981) corroborate this view and attribute fragmentation to adsorbed water alone.

Many minerals are ionically neutral, others have unsatisfied anions at their margins. Even if neutral, once in the groundwater zone, small amounts of water percolate into the lattice, taking some constituents into solution and thus creating ionic instability. Thus orthoclase, $KAlSi_3O_8$, is stable when fresh, but if the K atom is leached, some oxygen anions are unsatisfied and could attract water molecules. Slightly altered minerals may thus attract adsorbed water. The intercrystal water layers cause disaggregation without alteration. The water also neutralises ionic bonding between minerals and, hence, further weakens the rock. Hydration shattering may thus go some way to explain why some apparently unaltered rocks are nevertheless disaggregated. Once rendered permeable by physical shattering they are more readily infiltrated by water and some or all of the constituent minerals react with water and are chemically changed.

D. FACTORS INFLUENCING THE WEATHERING OF GRANITE

(i) Climate

As expected, most of the deep weathering so far recorded from granite terrains derives from the humid tropics, for there is there, of course, an abundance of moisture and most chemical reactions take place more rapidly at high than they do at low temperatures; As Agassiz (1865) put it 'warm rains falling on the heated soil must have a very powerful action in accelerating the decomposition of rocks.' In southeastern Brazil, in the Rio de Janeiro area, the biotite of fresh granite used in public buildings begins to show signs of decay after only thirty years of exposure to the elements, and in Malagassy (Madagascar) after only a decade. The efficacy of weathering in such environments is reflected in the depth of weathering recorded: more than 200 m in parts of southeastern Brazil (Barbier, 1957) and up to 100 m elsewhere in that country (Branner, 1896), 50 m in Nigeria (Thomas, 1966), and so on. But these profiles, like the 120 m depths of weathering recorded from eastern Australia (Ollier, 1965), the 40-50 m thick regoliths located in many parts of semiarid northern Eyre Peninsula, and the 40 m recorded from the southwest of Western Australia (Berliat, 1965), may reflect either long duration of weathering or past periods of humid warm climates, as much as present conditions.

The converse argument, of course, is that in arid conditions granite is resistant to weathering and in detail the suggestion finds considerable support (Barton, 1916; Bain, 1923).

In cold lands, on the other hand, conditions are evidently different, for, whether frost riving or hydration shattering is responsible, granite has commonly been split into slabs rather than being deeply altered, though it is arguable that in many areas any preglacial regoliths have been evacuated, as a result of erosion by late Cainozoic glaciers.

Whatever the details of the process, the distinctively shattered appearance of granite outcrops in cold lands, the characteristic talus cones and aprons (Fig. 3.11) surely attest the effectiveness of frost riving. Fracture zones are susceptible and the capacity of the rock or regolith to hold water is also significant (Godard, 1979) but, all in all, frost shattering possibly combined with hydration shattering appears to be one of the more effective means of achieving the disintegration of granitic rocks.

Fig. 3.11. Granite cliffs with talus cones at base, Andorran Pyrenees.

Some writers, however, and notably R. Dahl (1966), have adduced evidence to suggest that chemical alteration is also active in high latitude regions such as northern Scandinavia. Certainly grus containing altered minerals is present adjacent to fractures (see e.g. Watts, 1979), though some authors, and particularly E. Dahl (1961, 1963), attribute such developments to the warmer climates of interglacial or preglacial times. If the grus is not inherited then there is the implication of rapid chemical alteration, for weathering has proceeded at a pace sufficient to produce detectable change in what are, in some instances, quite youthful glacial tills.

Rapid weathering, possibly by a combination of chemical and physical (freeze-thaw) attack, is confirmed by the reported increases in the diameter of rock basins (see Chapter 8) in Bohemia by 1-3 cm in a decade and 1-2 cm in 25 years (Czudek *et al.*, 1964).

(ii) Rock Composition

Whatever the precise nature of the processes, many workers have concluded that composition is an important factor determining the nature and the rate of rock disintegration and decomposition.

Goldich (1938) pointed out that the susceptibility of the common rock-forming minerals to chemical weathering is the reverse of the order in which they crystallise from an igneous melt, because the high temperature minerals are in greater disequilibrium than those that crystallise in cooler conditions:

TABLE 3.3

Order of crystallisation from an igneous melt (Goldich, 1938).

Several writers implicitly corroborate this general scheme in respect of granite decomposition, and in the field Goldich's inferences find ready though only general support in the relatively resistant behaviour of granitic compared with other rocks. Even in conditions conducive to the rapid disintegration of granite, masses of that rock have obviously withstood weathering and erosion better than have other rock types, for they remain upstanding.

Thus, in Encounter Bay in South Australia (Fig. 3.12) - so-called because of the meeting thereabouts of the explorers Baudin and Flinders in 1802 - the granite, despite its coarseness, despite the abundance of orthoclase, despite its general susceptibility to alteration under attack by moisture and particularly by alkaline waters, nevertheless forms islands and hills because the biotite-rich schists of the Cambrian Kanmantoo Group into which the granite was emplaced have been weathered and eroded much more rapidly. Similarly, in the southwest of England it is the granite massifs that underlie Dartmoor, Bodmin Moor, Lundy and the Scilly Isles that form uplands and islands.

Fig. 3.12. Granite islands in Encounter Bay, S.A.

Granite is one of the more resistant of the common rock types. But granites vary in composition and this plays some part in determining the relative toughness of the specific types of granitic rock. Birot (1950), like Isherwood and Street (1976), placed particular emphasis on biotite as a source of weakness. Working in French Guyana, Hurault (1963) reached similar conclusions, and Dumanowski (1964) pointed out that though the granitic rocks of the Karkonosze Mountains of southern Poland vary in composition, all are comparatively resistant because they all lack biotite and other ferromagnesian minerals. Lamego (1938) attributed the vertical western face of the Pão de Assuçar bordering the Bahia Guanabara, partly to the presence of a bed of biotite gneiss that is much more readily weathered than the lenticular gneiss of which the main mass of the residual is built (Fig. 13 (a)). In the same region the eye-catching castellated summit of the Gavea consists of a mass of granulitic rock resting on lenticular gneiss which constitutes the main mass of the residual (Fig. 3.13 (b)). On the other hand, the spectacular peak of Corcovado (Fig. 3.14) is capped by lenticular gneiss standing on a broader base of granulite.

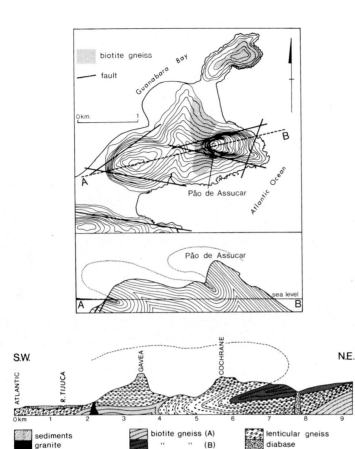

Fig. 3.13. (a) Above, map and section through the Pão de Assuçar showing steepening of western face through preferential weathering of bed of biotite gneiss. (b) Section through the Gavea, near Rio de Janeiro, which stands 842 m above sealevel and consists of a castellated summit mass of leptinolitic gneiss surmounting a dome of lenticular gneiss. Known as the Cabeça do Imperador (the Emperor's Castle), it is believed by some Brazilians to have been carved by the Phoenicians to represent Badzer, King of Tyro (after Lamego, 1938).

Klaer's (1956) experience in Corsica led him to stress the importance of variations in rock composition over all other factors. Thus, in the Punta di Mantelluccio and Punta Secca the granite with significant amounts of ferro-magnesian minerals (fayalite and hastingsite) forms vertical cliffs (Fig. 3.15) that contrast with the more gentle, though still steep, slopes developed on granodiorite and monzonite (Durand Delga, 1978). Williams (1968) reports unweathered pegmatite surrounded by weathered biotite-gneiss beneath the unconformity between Torridonian sandstone and Lewisian rocks in northwest Scotland, and so on - such correlations between rock type and surface form are

Fig. 3.14. The peak of Corcovado - the Hunchback - standing 704 m above sealevel, consists of a capping of gneiss standing on a granulite base within the
metropolitan district of Rio de Janeiro (Inst. Bras. Geograf.)

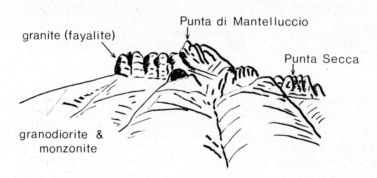

Fig. 3.15. Sketch of the massif called Mantelluccio, Corsica, seen from the
northeast (after Durand Delga, 1978).

commonplace. Lagasquie (1978) has suggested that the ratios of quartz,
alkaline feldspar and plagioclase (Fig. 3.16) determine the relative resistance
of any given rock, and illustrates his argument with examples from the Massif
Central and the Pyrenees. In these terms granodiorite is one of the weakest of
the granitic rocks. But other factors come into play. In the Corsican example
cited granodiorite is apparently more readily weathered than other varieties of
rock that are richer in ferromagnesians and that ought, on that account, to be

more vulnerable. Again, in the Everard Range (Fig.1.9) dolerite sills (see Jack, 1915) are more resistant than some parts of the granitic mass, though the ridges they form do not stand as high as the main granitic uplands.

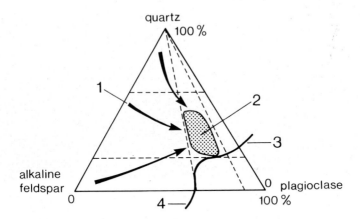

Fig. 3.16. Comparative resistance of plutonic rocks in the Pyrenees (after Lagasquie, 1978).

Notwithstanding such anomalies, varied susceptibility to moisture attack on the part of the essential granite minerals have several results. One such result, a minor one but of some practical significance and well illustrating the importance of composition in rock weathering, is that quartz is commonly left upstanding on granite surfaces subject to moisture attack. Such microrelief has been called pitting (Twidale and Bourne, 1976a), though it should be noted that Branner (1913) used the term 'pitted' to denote the development of what are here called rock basins on granitic exposures, and Jahn (1974) amongst others also refers to such basins as 'pits'.

Pitting is the roughening of the surface of the granite (Fig. 3.17) or other crystalline rock due to differential weathering of the constituent rock-forming minerals (Fig. 3.18). It is a micromorphological feature developed at the crystal scale. Usually quartz remains in relief as a result of the weathering of feldspar and mica, though feldspar phenocrysts also tend to resist weathering and remain upstanding. The form is widely developed, having been observed in areas as varied as the Nile Valley (Barton, 1916), Rio Province in southeastern Brazil, the Snowy Mountains of N.S.W., the deserts of central Australia, Zimbabwe, Namibia, several parts of South Africa, tropical humid Australia, Malaysia and Hong Kong (Figs. 3.5 and 3.19).

In gneissic rocks the pitting develops in linear patterns following gneissosity. In coarsely crystalline rocks, and especially in porphyritic

Fig. 3.17. Differential weathering by soil moisture at the weathering front and at the crystal scale causes the rock surface to develop a rough or pitted appearance in detail. Such rough surfaces stand in contrast with the smooth surface of long-exposed rock, as seen here the northern slope of Pildappa Rock, northwestern Eyre Peninsula, S.A., where the soil has been cleared from the footslope in connection with a water conservation scheme (see Twidale and Smith, 1971).

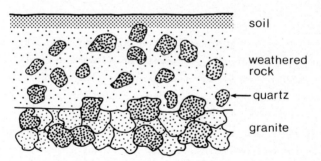

Fig. 3.18. Pitting developed at the weathering front.

granite, the pitting may take the form of a general roughening rather than clearly defined patterns of miniature relief, but, in either case, the differential weathering undoubtedly forms at the weathering from beneath the soil cover: eyewitnesses describe pitting as present on newly exposed rock surfaces, and one can strip away the soil cover to reveal pitted surfaces on the weathering front. How long pitting persists after exposure doubtless varies from site to site according to climate, chemical character of soil moisture, drainage, and so on. It is eliminated by fire flaking, frost action, by flaking due to moisture attack and possibly to insolation, and by the dissolution of upstanding quartz crystals by water trapped in the micro depressions adjacent to them;

Fig. 3.19. Pitting at Mt Bundey, near Darwin, N.T.

but it has persisted for at least sixty years on northwestern Eyre Peninsula
with no obvious signs of having been significantly subdued (Twidale and Bourne,
1976a).

Pitting itself does not constitute an absolute dating method, for it indicates recency in a geological rather than an historical sense. But even with
that limitation, it is useful, for it indicates not only that recent erosion has
occurred but it also provides a measure of the amount of erosion that has taken
place.

Thus, in the Tampin area recent erosion of some 2 m is indicated. Many of
the residual boulders there are pitted from base to crest, albeit with patches
that are smooth where the roughened surface has been removed by fire flaking
(Fig. 3.5). In many cases the scales that have flaked away can be found at
the base of the boulder. Invariably one side of the flake is smooth and the
other, the original outer surface, is pitted. The subsurface origin of the
pitting finds corroboration in the flared sides of the boulders (see Chapter 9),
which suggests that many, if not all, of the residuals were formerly part of the
regolith and surrounded by grus.

(iii) Texture

Rock texture is another variable affecting the progress of weathering. Pro-
viding there is access to crystal faces in pores and intergranular spaces, fine-
grained rocks ought to be susceptible to chemical (moisture) attack because,
compared with coarse-textured materials, they possess large areas of crystal

surface per unit volume. These surfaces have high free energy and are prone to
reaction with circulating liquids. Some crystal boundaries are planes of weakness exploitable by moisture (Fig. 3.20) and, again, fine-grained rocks possess
large areas of such partings. Coarse-grained rocks, on the other hand, ought
to be relatively resistant by virtue of their lesser areas of crystal face per
unit volume.

Fig. 3.20. The importance of partings, no matter how fine, as avenues of water
infiltration is demonstrated by the excess weathering of the fine-grained matrix
of this porphyritic rock exposed on the Leeukop, a bornhardt near Potchefstroom
in the western Transvaal, R.S.A. Water has penetrated along the boundary of the
porphyroblastic feldspar and has altered the adjacent rock so that the larger
crystals are left in slight relief due to the development of miniature depressions around them.

Such textural factors are clearly responsible for landform patterns in some
areas. Thus, Dumanowski (1968) reported examples suggesting that fine-grained
rocks are more resistant than coarse ones from such climatically contrasted
regions as southern China and the Karkonosze Mountains of southern Poland. In
South Australia the dacitic-rhyolitic suite of the Gawler Ranges is certainly
more resistant than the adjacent and essentially cosanguinous granitic rocks of
Eyre Peninsula.

On the granite domes south of Karibib, central Namibia, pegmatite sills and
veins are much more susceptible to weathering than the country rock, demonstrating the significance of grain size as a factor in weathering (Fig. 3.21).

Fig. 3.21. This pegmatite, which occurs as a sill intruding granite gneiss near
Karibib in central Namibia, is clearly less resistant than the host rock.

Similar pegmatite sills also illustrate the importance of relativity, for,
whereas the pegmatites are weaker than granite and form shallow clefts where
they intrude this rock, on nearby hills, where similar pegmatitic veins intrude
schists, they form low linear ridges.

 Branner (1896) also noted that in one area in Brazil coarse granite is more
readily weathered than is fine-grained rock of similar mineralogy, but that in
another locality there are prominent peaks underlain by coarse-grained rocks.
The answer may be found in fracture density. Again, though in apparent contrast,
on Dartmoor, in southwestern England, the fine-grained Blue Granite is, in some
circumstances at least, more vulnerable to weathering attack than is the very
coarse and strongly porphyritic Giant Granite (Brammall, 1926; Brammall and
Harwood, 1932) which is exposed over most of the upland and which forms most of
the well-known tors (Fig. 3.22). Closer jointing may again explain the apparent
anomaly (Waters, 1964), though the higher biotite content of the Blue Granite
and the possibility of soil moisture attack when the level of the high plain
was slightly higher, are also to be borne in mind.

Fig. 3.22. The coarse porphyritic Giant Granite that is exposed over most of Dartmoor makes up the greater part of Haytor West. A finer grained Blue Granite is exposed at the base of the low bounding bluff and is marked by the development of a shallow shelter. This could be due either to closer jointing, higher biotite content or soil moisture attack, or to a combination of these factors.

(iv) Partings

Not all variations in the form of the land surface developed on granitic rocks are comprehensible in terms of mineralogical or textural variations, and this has led many workers to emphasise yet another factor: namely rock fractures or partings. Thus, Lautansach (1950) concluded that in the Iberian Peninsula and Korea, composition is less significant in determining landform development than rock texture and fracture density. Godard (1979) noted that in northern Labrador severe frost shattering is restricted to shear zones and other areas of dense fractures. Birot (1952, p. 301) emphasised the role of rock fissures and fractures:

Cependant, le facteur le plus important de l'érosion differentielle parâit être la densité de diaclases et la multiplicité des lignes de fracture.

Any parting is a potential avenue for water infiltration, and is hence a source of weakness (Fig. 3.20). Crystal cleavage, intracrystal dislocations and crystal boundaries are avenues penetrable by water, but, of course, the essentially regular patterns of joints and faults are more important partly

because they are frequently open, partly because they are so widely developed,
and partly because they tend to form continuous networks (though it must be
added at once that many faults and joints are unpredictably and inexplicably
discontinuous). But fractures, of whatever origin, are planes of weakness,
that have been exploited both by molten materials from deep in the earth's
crust, and by external agencies, notably meteoric waters.

Many fractures have been intruded by hot gases, liquids and rocks migrating
upwards towards the land surface. In some instances, the erstwhile partings
have been sealed and are now difficult to detect. For instance, in the western
Murray Basin of South Australia, the Black Hill norite is subdivided by
orthogonal joint sets, but within many of the cubic and quadrangular blocks so
defined there are planar zones rich in chlorite that are interpreted as members
of an earlier generation of joints that have been welded as a result of later
mineralisation. At Mt Monster, a residual of porphyritic granite near Keith,
in the southeast of South Australia, some joints are open but others are partly
sealed by hydrothermal minerals including epidote and fluorite.

More commonly, the fractures are widened and filled by material that stands
in sharp mineralogical and textural contrast with the country rock. Some such
veins are thin, discontinuous and diffuse, and their morphological effects are
so minor that Spanish workers (e.g. Marti Boni and Vidal Romani, 1981) refer to
them as *nervaciones* - nerves (Fig. 3.23). Yet others, more substantial

Fig. 3.23. This granite in the Andorran Pyrenees has been intruded by many
fine veins and filaments of quartz which are more resistant than the host rock
and therefore form low ribs.

(Fig. 3.24), have nevertheless been planed off flush with the general surface (see also Debon, 1972, his Pl. 11.5). Others are weaker than the surrounding rock and form slots or clefts (see e.g. Shaler, 1887-8, at his Pl. LVI). Some, however, stand out because they are more resistant and stand in relief; others are weaker than the country rock and stand in *intaglio*.

Fig. 3.24. Quartz veins planed off flush with surfaces of gneiss, Nanutarra, northwest of W.A.

On Paarlberg, in the western Cape Province, R.S.A., aplitic sills are up-standing, but the partings that separate the sills from the country rock have been weathered to form distinct, shallow depressions (cf Fig. 3.25) and similar assemblages occur at other sites. Miniature ridges of similar origin are also well-known.

Fig. 3.25. This aplitic sill on Paarlberg, near Cape Town, R.S.A., is more resistant than the granite immediately adjacent to it but less tough than the main mass of granite to either side. This reflects the penetration of moisture along the margins of the sill and the alteration of the granite with which it came into contact.

Fig. 3.26. (a) Orthogonal fractures have produced a pavement of granite blocks near Karibib in central Namibia. The partings are unusual in that they are bordered by thin, but distinct, rims. Note also slightly raised ridge of quartz. (b) Detail of raised rims, which, on close examination are found to be protected by thin skins of recrystallised rock developed by frictional heating of the rock during shearing.

In the Karibib area of central Namibia, fracture traces exposed on granite platforms are bordered by low but distinct rims which, on inspection, are found to be due to the fractures being faults, slickensides due to differential movement being clearly discernible on the fracture planes (Fig. 3.26); the partings

Fig. 3.27. Joint exploited by moisture attack on one of the Kwaterski Rocks, northwestern Eyre Peninsula.

Fig. 3.28. Paarlberg near Cape Town, R.S.A., where a minor cleft can be seen in the foreground and a major slot in the middle distance.

are lined by recrystallised rock that is more resistant than the host granite. Jeje (1973) illustrates similar raised rims bordering fractures on a platform in southwestern Nigeria but does not comment on their origin.

Far more commonly, however, fractures have been exploited by water penetrating downwards from the surface. One result is the formation of clefts or slots, some

Fig. 3.29. So-called crazy paving due to the weathering of vertical joints exposed on a granite platform in the Devil's Marbles, N.T. (Hammer gives scale)

tens of metres long and several metres wide, others short and narrow, some simple, some complex (Figs. 3.27 and 3.28). They are widely distributed and are found, for example, in the humid tropics, as well as in arid and semiarid regions.

Vertical or near-vertical fractures that have been so exploited by weathering are known as *Kluftkarren*. On pavements, the traces of such fractures outline crazy-paving patterns (Fig. 3.29), or 'mud cracks' (Lister, 1973).

Water is the most important single agent at work causing the weathering of granite. Solution is of utmost significance, opening the way as it does to further reactions, but the initial breakdown may, in part at least, be physical and involve adsorbed water. Though the significance of rock composition and texture is not to be underestimated, the field evidence strongly suggests that the pattern and condition of fractures together constitute the most important single factor determining landform development in granitic rocks. Indeed, it can, with reason, be argued that because of the characteristic low permeability of the crystalline rock the importance of fractures is enhanced. Fractures are virtually the only means whereby water can infiltrate the rock, and water and weathering thus are concentrated in these planate zones.

The influence of joints and joint patterns is next discussed in relation to one of the most characteristic of granite landforms, the boulder.

PART II

MAJOR FORMS AND ASSEMBLAGES

CHAPTER 4

BOULDERS

Standing either in isolation or in groups or clusters (Fig. 1.8) residual boulders are, apart from plains, perhaps the most common of all granite landforms. They are certainly the most numerous and widely distributed of the positive relief forms developed on granite. They range in diameter from about 25 cm as, for instance, in an equigranular, fine-grained grey granite developed on the Palmer stock in the eastern Mount Lofty Ranges in South Australia, to 11-23 m in the Devil's Marbles in the central part of the Northern Territory, and some 33 m in the huge ovoid boulder known as the Leviathan in the Mt Buffalo complex of northeastern Victoria (Fig. 4.1); but the modal diameter is between one and two metres. Boulders vary in shape from spherical to ellipsoidal and also in the degree of roundness attained. Some are virtually perfect spheres while others are almost cubic, only the corners and edges of the original blocks being rounded.

As is indicated by the examples cited here, boulders are found in many, if not all, climatic regimes, including recently deglaciated regions in northwestern Europe and Britain.

A. SUBSURFACE EXPLOITATION OF ORTHOGONAL FRACTURE SETS

As has previously been mentioned, granites are characteristically well-jointed and are in particular subdivided into cubic or quadrangular blocks by orthogonal sets.

In many parts of the world spherical masses of intrinsically fresh rock set in a matrix of weathered rock are exposed in cuttings, quarries and natural cliffs (Fig. 4.2). Wagner (1913) considered them to be due to contraction during cooling, but their essentially near-surface occurrence, a general diminution in degree of rounding and increase in size of spheroidal masses with depth (see below) suggest an origin related to processes active at or near the land surface, not to primary magmatic effects. They are widely interpreted as being due to differential weathering attack beneath the land surface. Meteorological waters percolate down and along joints (Fig. 4.3). The rock immediately adjacent to

Fig. 4.1. The Leviathan, northeastern Victoria, is a huge isolated granite boulder measuring 33 m x 21 m x 12 m (Geol. Surv. Victoria).

Fig. 4.2 (a). Corestones set in a matrix of grus and exposed in a road cutting near Lake Tahoe, eastern Sierra Nevada, California.

the fractures is attacked by the moisture. As weathering progresses, the outline of the fresh rock mass changes from angular to rounded because of what MacCulloch (1814, pp. 71-72) described as 'more rapid disintegration at the angles than at the sides'. Logan (1849, p. 24) expressed the same thought:

That an angular block must disintegrate more rapidly on the edges than elsewhere is evident, for there are two faces, both subjected to meteoric action approximate and meet so that the edge decomposes quite through . . .

Fig. 4.2 (b). Corestones set in a matrix of grus and exposed in a road cutting in the Snowy Mountains, N.S.W.

In this way, what Jones (1859) called the 'heart of the block', the kernel, floater, *boule*, rock-kernel, a block of hard granite (Jahn, 1974), core-stone, core-boulder or *corestone* (see Scrivenor, 1931, p. 136; de Martonne, 1925, pp. 631-634; Linton, 1955), is formed within each joint block, surrounded by a mass of weathered granite commonly referred to as *grus* (the gruss of some American workers). Some authors (e.g. Brunsden, 1964) differentiate between mechanically disintegrated grus and chemically altered growan, but there is a gradation rather than a sharp distinction to be made between the two and the

Fig. 4.2 (c). Corestones set in a matrix of grus and exposed near Gemencheh Quarry, West Malaysia, beneath a laterite profile.

term grus is commonly applied to granite sand or fine gravel, regardless of whether some of the constituent particles have suffered alteration.

The roundness of corestones is due to differential weathering. They do not vary in composition from the rest of the blocks; in particular, they are not inherently more resistant than the marginal areas, as implied by Jameson (1820), Kingsmill (1862) and later by Pirsson and Schuchert (1915, p. 28). That the complex of corestones and grus is *in situ* is proved by the presence of veins traversing the entire rock mass (Kingsmill, 1862; see also Fig. 4.4).

Corestones of intrinsically fresh granite set in a matrix of weathered, friable rock constitute an example of spheroidal weathering. They also represent the first stage in the development of boulders. Similar marginal weathering surrounding corestones has been noted in such basic igneous rocks as basalt (e.g. north Queensland and the Drakensberg of South Africa), gabbro (Giles Complex, northwest of South Australia), norite (Black Hill, western Murray Plains, South Australia - see Hutton *et al*, 1977) and so on. Rather surprisingly, these weathering assemblages are also developed in arenaceous rocks, as for instance in an impure sandstone in the Pichi Richi Pass, near Quorn, South Australia.

Fig. 4.2 (d). Corestones set in a matrix of grus and exposed at Mt Bundey, near Darwin, N.T.

Joint-controlled subsurface weathering transforms an essentially homogeneous rock mass into two contrasted types of material, namely the corestones of fresh rock and the grus matrix. Thus, as the land surface is lowered, the friable grus is readily washed (or, more rarely, blown) away, whereas the corestones are too massive to be moved and are left *in situ* (Fig. 4.2). In some areas (e.g. in the Valley of the Agout, in the Sidobre de Castre, southern France; in several valleys of the Tijuca Massif near Rio de Janeiro; and at Palmer, South Australia) a chaotic mass of boulders known in France as *compayrés* (de Martonne, 1925, p. 632) is left behind (Fig. 4.5). But whatever the cause, the exposure of the corestones by differential erosion of the differentially weathered granite is the second stage in the development of boulders. Once exposed to the air and in relief, the boulders are no longer in constant contact with moisture and, because granite is susceptible to moisture attack but stable when dry, they tend to persist (see e.g. Thomas, 1974b, p. 32, his Fig. 10).

Two distinct processes, weathering and erosion, are involved in the formation of boulders and they are frequently referred to as the *two stage* process or mechanism, but the two are not necessarily separate and distinct in time (Lewis,

Fig. 4.2 (e). Corestones set in a matrix of grus and exposed in the Rocky Mountains near Boulder, Colorado.

in discussion of Linton, 1955). If erosion outpaces weathering, the erstwhile corestones are exposed as boulders; but if weathering proceeds more rapidly than erosion all corestones located in the near surface zone are reduced to grus.

Many of the observed variations in shape and size of boulders reflect the original joint spacing, though duration and effectiveness of weathering are also significant. Variations in joint spacing give rise to quadrangular blocks and triaxial ellipsoids. Turrets and spires are developed where the elongation is vertical (Fig. 4.6). Horizontal elongation produces cheesewrings (Jones, 1859; cheese-rings according to Howchin, 1981, p. 225) which are so called because of their resemblance to flattish rounds of cheese (Fig. 4.7). Some blocks and boulders remain essentially *in situ*, others are precariously perched on other blocks or on platforms - the 'perched bowlders' reported from Massachusetts (Shaler, 1887-8). They were known as logging stones to early British workers (Fig. 4.8), the term being derived from the verb to log or rock: they are blocks and boulders so finely balanced that they rock at a touch. They are also referred to as loganstones, balancing rocks, balanced rocks or perched blocks (Fig. 4.9). Cottage loaves comprise two or more boulders perched one on the other (Fig. 4.10) from their similarity to old-fashioned multi-tiered bread

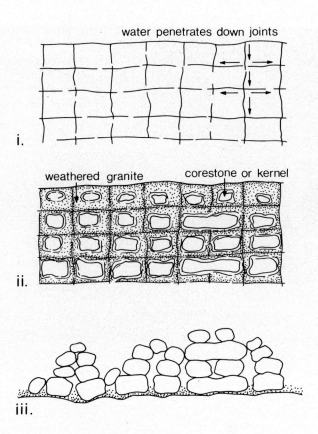

water penetrates down joints

i.

weathered granite   corestone or kernel

ii.

iii.

Fig. 4.3. Two stage development of boulders by differential fracture-controlled subsurface weathering and subsequent exposure of corestones by evacuation of friable weathered debris.

loaves. Gneissic layering (Fig. 4.11) gives rise to slabs, penitent rocks, monkstones or tombstones - the *Büssersteine* of Ackermann (1962). Boulders or groups said to resemble particular people, animals or objects are not uncommon (see, for example, Anderson, 1931; and Fig. 4.12).

The size of the boulders has a genetic implication, for, obviously, a corestone (and hence boulder) can be no larger than the original joint block, so that the diameter and spacing of juxtaposed corestones provide an indication of the minimum orthogonal joint spacing.

B. HISTORICAL PERSPECTIVE

The two stage theory of boulder development in granite is now widely accepted as the most satisfactory explanation for most, though not all, boulders (see below). Nowadays the concept is usually associated with the name of David Linton (1952, 1955) who subjected the field evidence to rigorous and critical

Fig. 4.4 (a). Corestones and grus exposed by the road side in the Rocky
Mountains near Boulder, Colorado, with a pegmatitic vein traversing both and
showing that the profile is due to weathering of igneous materials and not a
transported material such as glacial till.

examination, presented the argument with force, elegance and clarity, and later
defended the concept vigorously against all challenges (see e.g. Linton, 1964).
It is fair to say that Linton established the concept in the modern literature.

But the idea did not originate with him. Like many scientific concepts it
had a long gestation period; the two stage concept was appreciated in essence,
if not in name, at least a century, and perhaps 150 years, before Linton committed
the evidence and argument to print.

James Hutton is, in many ways, the father of modern historical geology. All
the great themes he discussed were firmly based in detailed field observations,
and he subjected even minor forms, including boulders, to close analysis.
Hutton recognised that some boulders are produced by attrition in rivers, in
glaciers or on beaches, and the term 'boulder' has to the present day in part
retained a connotation of being water-worn, or rounded by glacial action.
Throughout the last century the word boulder (or more commonly bowlder) was used
of worn and rounded masses of rock. Hutton had, however, much earlier recognised

Fig. 4.4 (b). Corestones and grus with veins *in situ* exposed in road cutting in northeastern Sardinia (R.F. Peel).

that some boulders may be formed 'by the decay of the rock around them' (Hutton, 1975, p. 174).

But Hutton, in turn, referred to the work of Hassenfratz (1791) in the southern Massif Central of France (near Aumont, between St Flour and Montpellier). There, some boulders and blocks sit half exposed on hillsides, while others stand one upon the other to form isolated pillars (Fig. 4.13). With remarkable astuteness and imagination, Hassenfratz realised that the exposed blocks are the more durable parts of the granite mass remaining after the rock around them had decayed and been washed away and that the partly and fully exposed masses of fresh rock represent stages in an evolutionary sequence. He noted that:

> on aperçoit tous les intermédiares entre un bloc de granit dur contenu et enchassé dans la masse totale du granit friable et un bloc entièrement degagé. (Hassenfratz, 1791, p.101).

At about the same time, Robert Jameson of Edinburgh reached similar conclusions about boulders he had observed on the island of Arran. He wrote that they:

> are not of very considerable size, and vary but little in that respect at the top or bottom of the glen, which shows that the greater part of them have not received their rounded form by attrition in the water of the glen, but are derived from a decomposed breccia. (Jameson, 1798, p.58).

Fig. 4.5. An episodic stream draining a small granite outcrop near Palmer, in the eastern Mount Lofty Ranges, has washed away all the grus in which the corestones were formerly embedded. The spherical masses have thus been exposed in a disordered linear group occupying the valley floor. In France such chaotic assemblages of boulders are known as *compayrés*.

Later, Jameson more accurately identified the 'breccia' from which originate the boulders characteristic of many granite outcrops, though he may not have appreciated that the matrix surrounding the buried spheres of fresh rock is merely weathered granite and not a different rock more susceptible to weathering:

> Some granites are disposed in rounded balls or concretions, which are from a foot to several fathoms in diameter .... The spaces between the concretions are filled with granite of a softer nature, which decays rapidly .... (Jameson, 1820, p. 414).

Jameson seems not fully to have grasped the part played by joints in the shaping of corestones and boulders, but in 1814 John MacCulloch noted that the granite of Devon and Cornwall is divided by fissures running mainly vertically and horizontally, so that the granite forms cubic and prismatic blocks (MacCulloch, 1814, pp. 70-71).

MacCulloch observed that 'Nature *mutat quadrata rotundis*', and noted that the blocks which, because of their higher elevation, have been exposed to weathering longer than those below them have become rounded. Granite balls are 'rendered spherical by decomposition' (MacCulloch, 1814, p. 76). MacCulloch was obviously

Fig. 4.6. Strongly developed vertical joints give rise to vertically elongate blocks that, when weathered, form tall turrets, in the Tassili of southern Algeria (P. Rognon).

thinking in terms of epigene attack, for he attributed them to longer exposure to 'air and weather' (MacCulloch, 1814, p. 71). Nevertheless, he discounted abrasion during transport as the reason for the rounding of the boulders and argued that the rocks are *in situ*.

MacCulloch also attributed the weathering of granite to solution by water, using the term solution in a general sense of chemical attack, though he also entertained the possibility of weathering being achieved by alternate wetting and drying (MacCulloch, 1814, pp. 72-74). He also noted that weathering can penetrate deep into rocks:

> a change may always be observed to have taken place from the surface downwards to a more or less considerable depth in the stone. Sometimes even the whole mass of rock will appear to have undergone this gangrenous process at once, and to have become a bed of clay and gravel. (MacCulloch, 1814, p. 72).

Some of the themes pursued by MacCulloch were developed by the great Henry de la Beche, who wrote of the granites of southwestern England thus:

Fig. 4.7. (a) Sketch of the Cheesewring, near Liskeard, Cornwall (after Jones, 1859). (b) Strongly developed horizontal joints give rise to horizontally elongate, slab-like blocks or cheesewrings. These examples are part of the Devil's Marbles, in central Australia.

> The granites are separated by divisional planes ... into cuboidal or prismatic bodies and the decomposition on the faces of these bodies, when the blocks are detached, and the superior facility for disintegration afforded by the corners would appear sufficient to produce the rounded character we often observe them to possess. (de la Beche, 1839, p. 450).

In the 1839 report, part of which has already been quoted, de la Beche follows MacCulloch and writes of decomposition affecting the faces of the joint blocks 'when the blocks are detached'. But he later presented a diagrammatic section

Fig. 4.8. The Logging Stone, St Levin's, Cornwall (after Jones, 1859).

(Fig. 4.14) showing rounded, ellipsoidal corestones embedded in weathered
granite (de la Beche, 1853).

In his later writings de la Beche may also have been influenced by two papers
due to Logan (1849, 1851), who travelled extensively in Malaya and Singapore
between 1830 and 1849. Logan's most perceptive observations concern Palo Ubin,
at the western end of the Johore Strait. In his earlier paper Logan saw the
boulder-strewn slopes of the island in unmistakably structural terms. He recognised the all-important part played by joints in determining the focus of
weathering:

> The blocks, protruding from the hills or ranges along the shores of Palo
> Ubin, are more solid and less composable masses and nuclei, of which the
> forms, with the directions of the sides and axes, have, in almost every
> instance, been determined by structural planes, and which remain after
> the surrounding rocks have been disintegrated and washed away. (Logan,
> 1849, p. 40).

He also reached the same conclusion as MacCulloch and de la Beche concerning the
origin of the spheroidal masses. He stated that the weakened edge will be
wasted away after every rain, so that, again like MacCulloch and de la Beche,
he obviously considered the possibility of rounding after exposure. But Logan
immediately challenged his own suggestion by stating, albeit with some licence,
that the globular blocks can be seen at Palo Ubin 'in the very act of separating
from the original compact in which they had been formed' (Logan, 1849, p. 24).

In his second paper Logan reinforced the suggestion that boulders originate
beneath the land surface by noting that on the coast of Palo Ubin (Fig. 4.15)
the seas have washed away the weathered rock:

> thus leaving the more resistant masses to emerge from the soil and stand
> out above the influence of decomposition. (Logan, 1851, p. 326).

When he mentioned that the boulders stand 'above the influence of decomposition', Logan was implying some contrast in weathering between exposed and buried
conditions, for he follows it with a footnote to the effect that:

Fig. 4.9 (a). At some sites the erosion of grus has left boulders precariously perched or balanced on platforms or on other boulders: Balancing Rocks, near Salisbury (Harare), Zimbabwe.

> When an exposed rock is attacked, the decomposing portion is washed or falls off, and the decomposition is arrested for the time. Under-ground decomposition tends to spread unchecked on all sides. (Logan, 1851, p. 326).

This may be the first reference to granite's peculiar property of being vulnerable to attack in moist conditions, but resistant when dry.

Thus, albeit in tentative fashion, Logan had, by the middle of the Nineteenth Century, grasped and expounded the essentials of boulder development through joint-controlled subsurface weathering.

The concepts developed by Hassenfratz, MacCulloch, de la Beche and Logan have never really been lost; they have merely been neglected from time to time. Nevertheless, the various component parts of the two stage concept can be traced

Fig. 4.9 (b). Perched boulder, Dartmoor.

Fig. 4.9 (c). Perched boulder in the Llano of central Texas, the feature being appropriately and hopefully known as Balanced Rock.

Fig. 4.9 (d). A perched boulder, the well-known Peyro Clabado, in the Sidobre of southern France.

perpetuated and, in one instance, developed from the eighteen fifties to Linton and the other moderns.

The Dartmoor granite and associated landforms were described in 1859 by T.R. Jones, who mentioned *granit pourri*, the decomposed granite nowadays known as grus, and also emphasised joint control of water infiltration and hence weathering:

> Granite ... is always cut through by joints or fissures .... The rain-water trickles through these lines of joints, decomposes the granite along the cracks, widening them and rounding off the angles of their intersections, and ultimately only the harder masses, or the hearts of the blocks defined by the joints remain as solid, crystalline granite. (Jones, 1859, p. 307).

At about the same time, Kingsmill (1862, p. 2) noted deep weathering of granite exposed on Hong Kong island and remarked on 'nodules' embedded in a

Fig. 4.10. A cottage loaf form at the Devil's Marbles, central Australia.

'soft yielding matrix'. He attributed the relative resistance or toughness of
the spheres to their being more quartzose than the surrounding rock, but demonstrated that the weathered mass was *in situ* by noting that the:

> original quartz veins of the granite, broken into small fragments by the
> forces which have operated on the surrounding rock, still traverse the
> disintegrated rock in all directions. (Kingsmill, 1862, p. 2). (See
> also Fig. 4.4).

Similar evidence had earlier been adduced by Darwin (1846) who noted that in
weathered gneiss near Rio de Janeiro minerals retain their positions in folia
oriented in the direction usual for the district. From mid century onwards many
writers referred either explicitly or tacitly to what are now known as corestones
and grus. Indeed, they excited no particular attention and seem to have been

Fig. 4.11. Gneissic foliation is exploited by weathering to give slabs that are either upright or inclined, according to dip. They have been called *Büssersteine* by Ackermann (1962), monkstones or penitent rocks in English. These examples are from (a), top, the Tungkillo area of the eastern Mount Lofty Ranges and (b), bottom, the Reynolds Range of central Australia.

Fig. 4.12. Various peculiar shapes and resemblances to living persons have been noted and named in granite outcrops. Bowerman must have had a curious physiognomy to have this example from eastern Dartmoor called Bowerman's Nose.

treated as a matter of course.

Hartt (1870, pp. 69-70) described boulders of decomposition from the Province of Espirito Santo in Brazil. Pumpelly, describing granite exposures he had observed in central Asia, reported that:

> the disintegrating mass consists of the rounded cores of the blocks surrounded by the decomposition products of the rest of the mass. (Pumpelly, 1879).

Archibald Geikie (1886, p. 20) published field sketches showing fresh orthogonally jointed granite merging upwards into corestones and grus, and then into granite soil (Fig. 4.16).

Many workers writing at this time mentioned joints in connection with weathering and the development of granite boulders, from Stockbridge (1888, p. 103) to Skertchley (1893); Branner (1896), who described 'boulders of decomposition' in Brazil, noting that in places they are left perched on points and ledges; Shaler (1887-8) who described 'decomposition bowlders' from Cape Ann,

Fig. 4.13. It was at this site, or one like it, in the Aumont district of the Margeride, southern Massif Central, that in 1791 J-H. Hassenfratz made the conceptual leap linking the corestones he saw partly buried in grus in the natural embankment, with boulders scattered over the nearby plains.

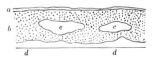

This illustration is taken from part of the road between Oke-hampton and Moreton Hampstead, Devon. *a* represents the vegetable soil; *b* decomposed granite; *c c* solid rounded masses of un-decomposed granite, included in the decomposed part; and *d d* solid granite.

Fig. 4.14. De la Beche's 1853 diagram of a corestone: (a) soil, (b) decomposed granite, (c) rounded masses of 'undecomposed granite' and (d) solid granite.

Massachusetts; Turner (1894) who illustrated spheroids surrounded by concentrically structured weathered rock in the Sierra Nevada; Reid *et al*. (1910) who recognised corestones; Pirsson and Schuchert (1915, p. 28) who mentioned 'boulders of decomposition'; de Martonne (1925, p. 631 *et seq*.) who referred to 'decomposition en boule'; and Larsen (1948, p. 114 *et seq*.) who described what

Fig. 4.15. J.R. Logan visited Palu (or Pulau) Ubin in the late eighteen forties and not only appreciated that the corestones set in grus he noted in the quarry faces were incipient boulders, but he also realised that the flutings so beautifully developed on the bare rock surfaces also had their origins in the subsurface (see Chapter 9).

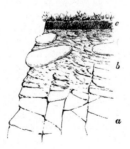

Fig. 4.16. Geikie's (1886) sketch of corestones within a weathered profile: a, solid granite; b, corestones in grus and c, soil.

he called 'boulders of disintegration' from the southern Sierra Nevada of California.

These and many others not only perpetuated the notion of corestones and of joint-controlled and differential subsurface weathering, but together they inadvertently demonstrated the essentially global character of the process.

C. TYPES OF PERIPHERAL WEATHERING

Differential weathering of joint blocks is widely developed but several variations in the type of marginal weathering have been described. Many corestones are set in a mass of grus, or granite that has been broken down by weathering processes to small fragments of sand or fine gravel (Fig. 4.2). Such weathering is known as granular disintegration. Some corestones are surrounded by layer upon layer of thin (1-5 mm) discontinuous flakes, slivers or laminae, which look to have been, as it were, wrapped around the corestone (Fig. 3.10).

At other sites the concentric layers are thicker (10-30 cm) and look like the leaves of an onion (Fig. 4.17); hence the appellation onion-skin weathering, which is also referred to as spalling. Finally, in some localities corestones have evidently been formed through the separation, by fracturing, of tetrahedral masses, each with a concave inner face, at each corner of the joint block. Some of the roughly ovoid core masses have flat ends and look like barrels. Such tetrahedral cornerstones associated with ovoid corestones have been observed in quarries at Palmer and on Granite Island, both in southern South Australia, and at Albany in the southwest of Western Australia.

Whatever the type of marginal weathering, however, almost all observers are agreed that the transition from the fresh rock preserved in the corestone to the friable, altered marginal areas is remarkably abrupt. There are physical contrasts between friable grus and cohesive crystalline rock, chemical differences between more-or-less altered feldspar and mica and fresh minerals. Although Ollier (1960, p. 45) states that the reason for this abrupt junction is not known, it is almost certainly due to the physical character of granite which is of very low porosity and permeability when fresh and cohesive but which becomes much more permeable once it is even slightly weathered (Kessler *et al.*, 1940).

Some workers have invoked insolation as the cause of disintegration and spalling, but such flaking is rare, superficial and discontinuous. Corestones and grus or other marginal weathering effects are commonly found up to 10 m beneath the land surface and in places they are known from depths of several scores of metres. Flaking and spalling occur as masses of laminated material extending around corestones, and the tetrahedral cornerstones also occur well below the ground surface. Thus, heating and cooling, even if the latter is aided by rain showers, does not explain the depths to which such marginal weathering extends.

Several workers have suggested that the concentric structure observed in the

112

Fig. 4.17. Thick concentric shells or layers known as onion weathering,from
their similarity to the leaves of an onion, Snowy Mountains region of N.S.W.

marginal areas of many granite joint blocks is due to the release of pressure
consequent upon erosional offloading (see also Chapter 5). Farmin (1937), for
instance, invokes the mechanism in explanation of onion weathering. He argues
that during cooling and crystallisation the granite becomes stabilised in an
environment characterised by high pressure, if only because of the weight of
superincumbent strata. That the overlying rocks have been eroded away is
evidenced by the very exposure of the granite, so that the vertical loading has
undoubtedly decreased in time. Farmin summarises his interpretation:

> the unloaded rock will tend to expand and will do so by fracture wherever
> internal stress exceeds its elastic strength. Not all the fractures will
> be concentric, but a concentric exfoliation is the ideal form of relief of
> the stress in a homogeneous rock. (Farmin, 1937, p. 632).

The argument is persuasive and both laboratory work (Bridgman, 1938) and practical experience in deep mines (Isaacson, 1957; Talobré, 1957; Leeman, 1962)
suggest that fractures parallel to the surface, whether natural or artificial,
develop under conditions of diminished lithostatic pressure. In deep gold mines
in South Africa, for example, an intradosal zone of tightly packed rock slivers
and blocks is commonly found around excavations (Fig. 4.18). It is stable
because the fragments are interlocking and because of frictional resistance.
The fracture zone is a manifestation of stress release,and, because it is in

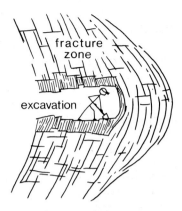

Fig. 4.18. Shattered intradosal zone developed around tunnel or other excavation (after Leeman, 1962).

equilibrium, rock bursts are relatively rare in this zone.

Yet against the suggestion of pressure release are the occurrence of concentric structures around corestones in such rocks as basalts that have never been deeply buried, the occurrence of the structure all round the corestone rather than preferentially on the upper sides, and the development of flaking on the interiors of tafoni located within joint blocks and sheet structure, both in granite and in such sedimentary rocks as quartzite (see Chapter 10).

Another explanation favoured by some is that the corestones reflect primary petrological structures. Jones (1859), for example, noted nodular or concentric structure in the Dartmoor granites, and Rondeau (1958), following several earlier workers, including Shaler (1869) and Ormerod (1869), attributed the formation of boulders to curved joints. Curved joints certainly exist (Fig. 4.19), but do not influence the shape of the boulders; primary sets of concentric or spherical fractures of a radius consistent with the observed size range of corestones and boulders have not been described from any locality.

On the other hand, it is apparent that the primary distribution of various minerals may have contributed to the development of corestones by influencing the course of weathering within joint blocks. Near the Tooma Dam Site in the Snowy Mountains, N.S.W., for example, mineral banding, a primary petrogenic feature, occurs in the marginal areas of blocks of diorite and the shape and size of the corestones is clearly related to these (Fig. 4.20). Again, in the Lake Chad region of central Africa, corestones of granite embedded in rhyolite are explained as a magmatic feature, with globules of still liquid granite having been mixed with the faster crystallising, and more easily weathered, rhyolite to form corestones and, in due course, boulders (Barbeau and Gèze, 1957 - see Fig. 4.21).

Duffaut (1957) attributed the preferential weathering of the marginal zones

114

Fig. 4.19. Curved joints exposed in shore platform at Smooth Pool, near Streaky Bay, Eyre Peninsula, S.A. (see also Fig. 1.7).

to microfissures and, though he offers no reason for its development, it could be due to fracture propagation caused by shearing along pre-existing fractures (see Twidale, 1980b; also Chapter 5). Similarly, the tetrahedral cornerstones described from some few sites cannot be explained in terms of insolation, pressure release or chemical attack, but are comprehensible in terms of the rotational shearing of pre-existing orthogonal joint blocks. Forms of spherical shape offer least resistance to such stresses, so that the elongate barrels

Fig. 4.20. Mineral banding in diorite and developed in between corestones, near Tooma Dam, Snowy Mountains, N.S.W. The hammer gives scale.

could have developed within joint blocks as a result of shearing (Fig. 4.22). Such a mechanism accounts for the observed forms, the parallelism between the long axes of the corestones and the regional tectonic style, and the notable absence of chemical alteration in the fracture zones.

Palmer and Nielson (1962) have suggested that the marginal rotting of joint blocks in the southwest of England is due to hot fluids and gases penetrating along joints and effecting hydrothermal metamorphism. They point to the survival of residual hills in the more massive, less susceptible, regions, though this distributional pattern is as readily explained in terms of resistance to meteoric water circulation.

The most commonly accepted and plausible explanation for much, though probably not all, of the marginal weathering developed around corestones is that it is due

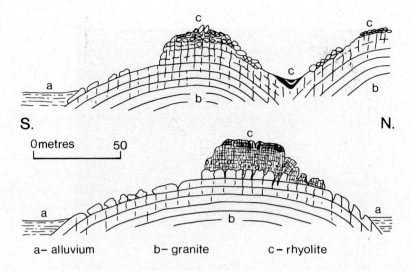

a– alluvium   b– granite   c – rhyolite

Fig. 4.21. Sections through inselbergs at Ngoura and Gamsous, Lake Chad region, central Africa (after Barbeau and Gèze, 1957).

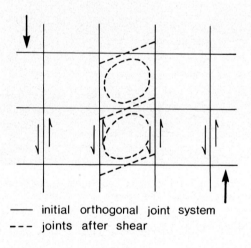

——— initial orthogonal joint system
--- joints after shear

Fig. 4.22. Development of barrel-shaped corestones and of tetrahedral corner-stones by the shearing of cubic blocks.

to chemical alteration caused by circulating groundwaters, and involving essentially alteration by solution, hydration, hydrolysis, carbonation and oxidation. Recent work (see Chapter 3; and Eggleton and Buseck, 1980) suggests that positive feedback effects stem from solution, crystal dislocation, further penetration of water, and so on. Moisture attack causes mineral changes and volume increase sufficient to rupture the rock.

In general terms, the intensity of weathering in the marginal areas increases outwards from the corestones (Chapman and Greenfield, 1949). This is consistent

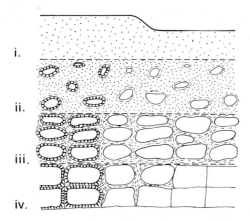

i.

ii.

iii.

iv.

Fig. 4.23. Weathering profile developed on granite and showing gradual decrease in intensity of alteration with depth: (i) grus; (ii) small corestones in grus matrix; (iii) large corestones (50-90 percent of whole) still recognisably derived from juxtaposed blocks, but with some intervening grus; (iv) rock with weathered debris along joints; solid rock more than 90 percent of whole but possibly iron stained (after Ruxton and Berry, 1957).

with circulating groundwaters having advanced from the outside toward the centre of each joint block, and the outer zones having been subjected to weathering longer than have the interiors of the blocks. For the same reason, the size of corestones ought to, and at many sites does (see Ruxton and Berry, 1957), increase with depth: the rock near the surface has been subjected to weathering longer than that at depth (Fig. 4.23). But there are many exceptions to this general rule (Fig. 4.24).

Some large apparently anomalous floaters doubtless reflect particularly wide fracture spacing, though it is not everywhere possible to demonstrate this because intense alteration can destroy all vestiges of the original structure. In drill holes and in vertical shafts it is frequently found that zones of fresh rock are underlain by rotted materials. This has been taken as evidence of hydrothermal activity, but rarely does the grus contain typical hydrothermal minerals and it is concluded that the apparent inversion merely reflects zones of contrasted fracture density, and therefore varied accessibility to circulating groundwaters.

Ruxton and Berry (1957) and others have shown that the granite is progressively weathered as the weathering front advances toward the centre of the joint block from the margins. The corestone is thus reduced in size and becomes more and more rounded (Fig. 4.23). At the weathering front the fresh rock is attacked by water and the biotite and feldspars are altered, releasing the iron oxides that give the reddish-brown tinge typical of such zones. The rock is converted to friable grus and then to a reddish-brown clay with fragments of quartz that

Fig. 4.24. In this 9-10 m deep road cutting in the Snowy Mountains, N.S.W. the grandiorite is deeply weathered, yet a large corestone persists quite close to the land surface.

suffer slow solution.

As water penetrates along the joints volume increase consequent upon alteration could cause the affected outer zone to separate from the main or host mass. As water penetrates further into each block, so more and more shells could be developed, but why some shells are thin (flakes) and others several centimetres thick is not clear. Presumably the contrast reflects the inherent tensile strength of the rock and the amount of volume increase (assuming there is an increase) induced by weathering. Also it is not clear why marginally weathered zones are concentrically structured while the main mass of weathered rock consists of structureless grus. It may be a function of texture, mineralogy and weathering, or alternatively it may be that in time and with increased intensity of weathering all flakes are broken down to grus.

Fresh granite is a remarkably strong rock. As mentioned previously, it consists of interlocking crystals and is additionally strengthened by inter-crystalline ionic bonding. Measurements made in the U.S.A. and Australia (d'Andrée *et al.*, 1965; Stapledon, 1961) show that the average tensile strength of unweathered granites and gneisses ranges between 70 and 140 kg/cm^2. Yet the laminae involved in the flaking around corestones display at most slight alteration of the feldspars and biotite, and it is difficult to visualise how such slight chemical alteration and production of hydrophilic clays could cause the rock to rupture, unless the adsorption of layered water has played a crucial part (see earlier, Chapter 3).

Again, if volume increase has taken place, and since various minerals are involved, the expansion ought to be differential and cause disruption of textures in the rock. Yet such features as mineral banding (Fig. 4.20), lineation and foliation remain undisturbed. Also, if water penetration is the cause of fracturing in the marginal zones, the flakes ought, in detail, to conform to crystal boundaries, whereas in fact they cut across such boundaries as well as cleavage.

It is for these reasons that Ollier (1967) favours an origin of concentric structure involving no volume change. Following Kieslinger (1932) and Carl and Amstutz (1958), he suggests that chemical reactions of the *Liesegang* ring type, involving diffusion and periodic reprecipitation and resolution of salts, presumably causing fatigue and disintegration, may be responsible for the observed evidence.

But a more likely reason for the minimal alteration noted in some of the thoroughly disaggregated granite rocks is that hydration shattering (see Chapter 3) has occurred. Certainly, such a mechanism accounts for the observed facts, though of course once a rock has suffered disaggregation water can more readily penetrate the mass and alteration (solution and hydrolysis) take place.

D. EVACUATION OF DEBRIS

Whatever the morphology, composition and genesis of the marginal zones, the weathered granite may eventually be evacuated and the corestones exposed as boulders. The transportation of the weathered rock is largely the work of wash, rills and rivers. Wind may play some small part in arid and semiarid regions, and solifluction is significant in nival areas (Linton, 1955), but it is running water that is primarily responsible for the exposure of corestones.

E. DISINTEGRATION OF SHEET STRUCTURE

Some boulders result from the disintegration of massive arcuate sheets or slabs which are typical of bornhardts, and were mentioned in Chapter 2. These arcuate masses of rock are split by radial fan-joints (see e.g. Balk, 1937) or stress planes that are disposed at right angles to each other and which thus effectively subdivide the thick slabs into blocks, sets of which can be seen *in situ* in orderly arrangement (Figs. 4.25 and 4.26) on the flanks of inselbergs in

Fig. 4.25. The thick slabs of rock known as sheet structure disintegrate at or very near the land surface into blocks that are converted to boulders partly in the shallow subsurface, partly or wholly under attack after exposure. A thick slab, subdivided into cubic or quadrangular blocks, can be seen on the crest of a bare hill near Fort Trinquet, Mauritania. Some rounded blocks have already gravitated to the base of the slope (R.F. Peel).

many parts of the world (e.g. Enchanted Rock, Texas; Everard Range, central Australia; Namibia, northern and central Namaqualand). On some residuals such blocky remnants of disintegrated sheets are confined to the upper levels. Thus on Mt Wudinna and Ucontitchie Hill there are remnants of several sheets seemingly resting on a massive, intrinsically unweathered domical mass of rock.

Similar groups of boulders obviously derived from the breakdown of massive sheets and developed on norite have been described from the Bushveld region of the Transvaal (du Toit, 1939, p. 171). Such sets of blocks in close juxtaposition can be seen still masked by a veneer of grus and soil at sites on northwestern Eyre Peninsula and elsewhere (Figs. 4.27 and 4.28). The rocks are slightly rounded, clearly as a result of subsurface moisture attack, but others are still angular on exposure. Some sheets suffer disintegration while still beneath the land surface and the blocks are converted to corestones set in grus (Fig. 4.28). Others, however, have been rounded and yet others completely

Fig. 4.26. Rounded blocks, some *in situ* and derived from the disintegration of sheet structure, Namibia (J.A. Mabbutt).

broken down by epigene agencies (Fig. 4.29).

There are many variations in the pattern of weathering, but in some places (e.g. Little Wudinna Hill on northwestern Eyre Peninsula) the sheet remnants preserved on upper slopes consist of clusters of angular blocks, whereas those resting on lower slopes are rounded and distinguished by the development of tafoni and flared bounding slopes. Such variations presumably reflect a contrast between atmospheric and subsurface weathering attack, the latter being dominant on lower slopes, the former on the higher slopes and crests. It has also been noted, however, that comparatively few remnants remain on the present midslope, where gradients are sufficiently steep to cause blocks to slide down to lower levels, especially during rain.

Thus, though granite boulders are formed in several ways, many, probably most of them, are due to a combination of two processes involving, first, differential fracture-controlled subsurface moisture weathering which produces corestones set in a matrix of grus, and then streams and wash which evacuate the grus and expose the corestones as boulders. Such two stage development clearly applies

Fig. 4.27. Orderly row of subrounded boulders, partly set in grus, Little Wudinna Hill, northwestern Eyre Peninsula, S.A.

Fig. 4.28. Large, rounded blocks (x) exposed in quarry at Paarlberg, western Cape Province, R.S.A. sit in ordered inclined rows and are due to the subsurface weathering of sheet structure.

Fig. 4.29. This sheet structure on Little Rock in the Enchanted Rock complex of central Texas has broken down after exposure.

to granite boulders in many parts of the world, and has operated under various and varied climatic conditions. The concept, which can be traced back almost two hundred years to the writings of Hassenfratz, is now widely accepted as a rational and reasonable explanation for these, the most common of all residual forms developed on granitic rocks.

Boulders are not climatic indicators, as was assumed by some earlier workers, who mistook them for glacial erratics. Some, of course, are of this nature, in the sense that they have been transported, by glaciers possibly, being further rounded in the process. But most granite boulders are residual in that they remain after the grus and other weathered granite that originally enveloped them has been evacuated. Boulders can be considered to be convergent forms, or forms of equifinality in the terminology of some modern authors, for though most are the result of two stage development involving fracture-controlled subsurface weathering followed by differential erosion of the unevenly weathered mass, the precise nature of the weathering and erosional processes has varied from place to place and no doubt also from time to time at the same site.

CHAPTER 5

INSELBERGS

Inselbergs are ranges, ridges and isolated hills that stand abruptly from
the surrounding plains, like islands from the sea. Inselbergs are characterised
by steep bounding slopes which meet the adjacent plains in a sharp, almost
angular, junction. Inselbergs are of many shapes and sizes, depending largely
on their structure, but the granitic forms are of three major types (Figs. 1.10-
1.18).

By far the most common and widely distributed is the *bornhardt*, the domical
form named after the German geologist of that name who last century explored
East Africa, and provided some of the most evocative descriptions and beautiful
sketches of the forms and the landscapes of which they are part, as well as
astute analyses of their possible origins. Some bornhardts stand in isolation,
others occur in small groups, but in contrast with these detached froms there
are ranges or massifs that comprise ordered repetitions of the domical form.
Thus the Everard Range (Fig. 1.9), in the north of South Australia, and the
Kamiesberge of central Namaqualand (Fig. 1.10) each consists of a large number
of closely juxtaposed bornhardts arranged in ordered fashion. Each dome is developed on a massive joint block, but is nevertheless part of a larger whole.

Other granitic inselbergs are angular and castellated and are known as *castle
koppies*. Yet others are block- or boulder-strewn and are called *nubbins* (or
knolls). Neither is as frequently and widely developed as the domed variety,
which appears to be the basic form from which the others are derived (Twidale,
1981a).

A. BORNHARDT CHARACTERISTICS

Bornhardts are domical hills with bare rock exposed over most of the surface.
They are developed in massive bedrock in which open fractures are few. Typically
and widely developed in granite and granite gneiss, they are also formed in
other plutonic rocks as well as in sedimentary materials. For example, they
have been described from sandstone terrains in the Hombori region of Mali
(Michel Mainguet, 1972) and from central Australia, where Ayers Rock is a well-
known bornhardt formed in steeply-dipping arkosic sandstone (Gosse, 1874;
Ollier and Tuddenham, 1962; Bremer, 1965; Twidale, 1978b; Twidale and Bourne,
1978a). They are developed in conglomerate in the well-known Olgas complex of
central Australia, in Mt Bresnahan, in the western Pilbara of Western Australia,

and in the Mallos de Riglos, near Salinas, in the foothills of the Spanish
Pyrenees (Barrère, 1968). The conglomeratic towers of the Meteora region of
northern Greece are of similar character.

Orthogonal fracture sets are well developed in the granite preserved in
bornhardts. Indeed, the residuals are delineated by prominent vertical and near-
vertical fractures that form part of such sets (Fig. 5.1). Nevertheless, in
bornhardts they are subordinate to arcuate convex-upward sheeting joints (see
Chapter 2) which seemingly determine both the domical form of the residuals and
the shape of the bounding slopes or valley side slopes in hilly or mountainous
terrain (Fig. 5.2).

Finally, bornhardts are, and as noted by Obst (1923), Jessen (1936), King
(1949a) and others, characteristically developed in multicyclic landscapes, that
is, in landscapes in which remnants of palaeoplains preserved high in the relief
indicate former phases of baselevelling, subsequent relative uplift and stream
incision (Figs. 5.3 and 5.4).

B. REASONS FOR POSITIVE RELIEF

Several explanations have been offered for bornhardts standing above the
adjacent plains, but basically there are four main arguments, involving faulting,
lithological control, contrasts in fracture density, and cyclic development.

First, some few bornhardts are possibly of tectonic origin in that they may
be upfaulted blocks. Passarge (1895), for example, suggested that some of the
inselbergs he had examined in West Africa (Adamaua) were in fact minor horsts,
and similar conclusions have been reached in respect of certain bornhardts in
southeastern Brazil (Lamego, 1938; Barbier, 1957; Birot, 1958) and French
Guyana (Choubert, 1949). The Pic Parana is a well-known example of southeastern
Brazil (Fig. 5.5). But most bornhardts are not tectonic: there is no evidence
that most of them are delineated by large-displacement faults, and, even in those
examples cited as of tectonic origin, it is not everywhere apparent whether the
fault-delineated scarps are tectonic or structural, whether they are fault scarps
or fault-line scarps (see Twidale, 1971a, p. 109 *et seq.*).

Second, many authors interpret bornhardts as structural forms, that is, as
forms due to the exploitation of weaknesses in the crust by various external
agents of weathering and erosion. Thus there are two aspects to be considered -
the nature of the crustal variation and the character of the external forces.
With respect to the first of these, some workers have resorted to lithological
contrasts, involving for example variations in the mineralogy of the granites
and gneisses, as in respect of Stone Mountain, Georgia (Lester, 1938; Hermann,
1957). Others, like Thorp (1969) and Selby (1977), have drawn attention to
granitic uplands and adjacent schist plains in the Air Mountains (southern Sahara)
and southwestern Namibia respectively.

126

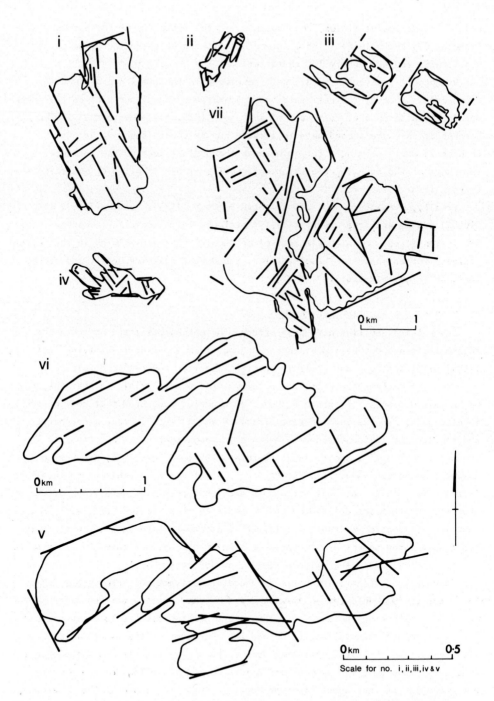

Fig. 5.1. Inselberg shape in plan related to fracture zones (i) Mt Wudinna, S.A., (ii) Yarwondutta Rock, S.A., (iii) Minnipa Hill, S.A., (iv) Pildappa Rock, S.A., (v) Hyden Rock, W.A., (vi) Mt Stirling, W.A., (vii) massif in central Namibia.

Fig. 5.2. This small fracture-controlled river valley near the Kylie Lake, in southern Zimbabwe, is flanked by slopes that are merely exposed convex-upward sheet structures.

Variations in rock composition and texture have probably been neglected as causes of differential weathering and erosion within plutonic bodies. As indicated in an earlier chapter (see e.g. Fig. 2.16), most plutons are complex composite features within which there are several bodies of varied petrological, and hence weathering and erosional, characteristics. Some of these internal variations are pronounced, others subtle. In some cases, including some of those cited, topographic and petrological patterns are closely coincident. Elsewhere, only general correlations are possible. For instance, at Stone Mountain, Georgia (U.S.A.), there is only a general coincidence between the upland and the outcrop of the Stone Mountain Granite on which it is developed, for plains as well as the dome occupy different parts of the exposure; on the other hand, the plains immediately to the southwest of the dome are eroded in a biotite gneiss

Fig. 5.3. The Bushman Kop, near Witrivier, in the eastern Transvaal, was a low bornhardt when streams were graded to the African Surface, remnants of which are preserved in the adjacent high plain (X). Subsequently, rivers were rejuvenated, further exposing the Kop and revealing several other domes.

(Fig. 5.6).

Finally in this context it must be pointed out that much of the evidence that would permit the significance of lithological contrasts to be assessed has been eroded. It is the character of the compartments of rock that were formerly located above the present plains, vis à vis the surviving bornhardt masses, that is crucial; and these have been eroded. All that can be done is to point to known compositional and textural variations within plutonic masses and to the possible geomorphological implications of such diversity.

Holmes and Wray (1912, 1913) suggested that some of the bornhardts of Mozambique are merely projections or apophyses developed at the margins of the plutons; du Toit (1939) made similar claims in respect of forms in Mozambique and Zimbabwe; and Worth (1953) interpreted the form of Dartmoor as broadly reflecting the shape of the pluton on which it is based. Again, a few residuals may be of this origin (Fig. 5.7), but in most areas there are indications of profound erosion of the crystalline rocks, suggesting that the original outlines of the plutons have long since been lost. In any case, and whatever the validity

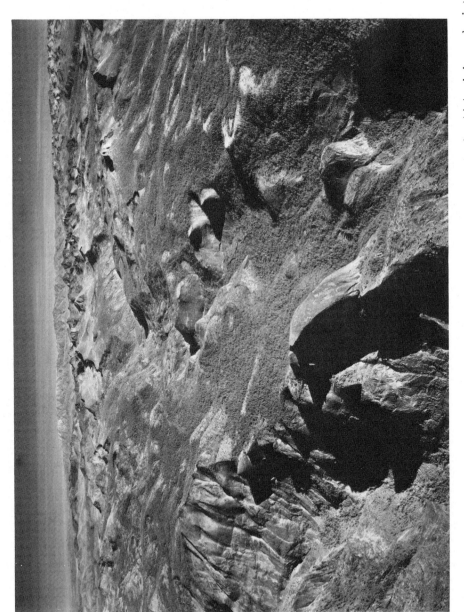

Fig. 5.4. Aerial view of the Yosemite region, Sierra Nevada, California, showing high plains, glaciated valley and various bornhardts, including Half Dome (U.S. Geol. Surv.).

130

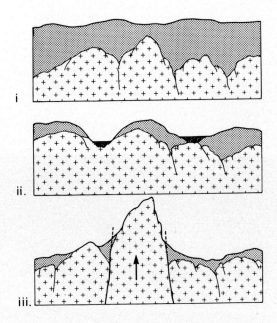

Fig. 5.5. The Pic Parana, in southeastern Brazil, is, according to Barbier (1957), an upfaulted mass of granite that has developed in the stages shown.

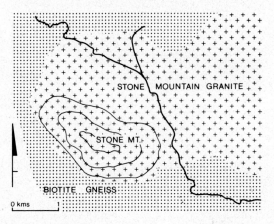

Fig. 5.6. Geological sketch of Stone Mountain and environs, Georgia, U.S.A. (after Herrmann, 1957).

of these claims, lithological contrast is again involved.

Many of the cases of lithological variation cited seem to be well substantiated, but in most instances the granite underlying the plains is mineralogically similar to, if not identical with, that of which the residuals are composed. For these, a substantial majority of bornhardts, the explanation that enjoys greatest support derives from observations suggesting that fracture density

Fig. 5.7. This small dome in central Namibia protrudes from a mass of schist and is clearly a small projection or stock.

varies from one compartment to another within a granitic mass. The theses of Lamego (1938 - see also Valverde, 1968), who interpreted the *morros* of the Rio de Janeiro region as the compressional cores of folded structures in gneissic rocks (Fig. 5.8), and Brajnikov (1953), who, working in the same general area, suggested that the residuals are essentially gigantic corestones or floaters developed as a result of metasomatic processes that caused volume increase and hence compression, are variations on this theme.

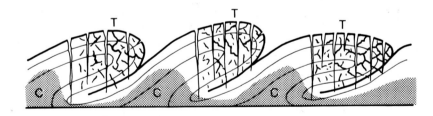

Fig. 5.8. Cross-section of the Rio area showing zones of compression (C) and tension (T) developed in the folded crystalline sequence, and with the future *morros* coincident with the compressional zones (after Lamego, 1938).

In more general terms, many writers have alluded to variations in fracture density as a significant factor in landform development on granitic rocks. Thus Le Conte (1873, p.327) noted that the massive domes of the Yosemite 'consist of hard material, little affected by joints' and, more specifically, Mennell, writing of the Matopos of Zimbabwe, stated:

the influence of the divisional planes of the rocks must not be overlooked, and it is to the variations in the number and character of the joints that the varied scenic aspects of the Matopos may be traced. Where stretches of comparatively level country occur, it will generally be found that the joints are numerous and irregular in direction, so that the rock readily breaks up and presents a large surface to the agencies of disintegration. In such cases the superior hardness of particular bands avails them little, as they are unable to show a solid front to the disrupting forces. On the other hand, joints are often entirely absent over a considerable area, and the tendency of the rock then is to weather into smooth rounded surfaces with a very large radius of curvature. Probably the actual outlines of the hills or ridges and the general direction of most of the Matopo valleys are determined by master joints, occurring at long intervals, which have formed a starting point for the work of erosion. (Mennell, 1904, p. 74).

That there are contrasts in fracture density between hill and plain is clearly seen at such sites as Blackingstone Rock, on eastern Dartmoor, where the massive residual contrasts strongly with the well-jointed granite exposed in a quarry (the source of the granite from which London Bridge was taken; the bridge now spans Lake Havasu, in Arizona) excavated into the nearby high plain (Fig. 5.9). Again, in water storages excavated at the margins of Ucontitchie Hill (Fig. 5.10) the granite is seen to consist of corestones set in weathered granite. The spacing of the corestones (see Chapter 4) suggests that the original fracture spacing was 1-2 m, in contrast with the 20-30 m between fractures discernible on the nearby residual. Büdel (1977, p. 109) has illustrated a similar juxtaposition of fresh, massive rock in an inselberg and corestones in grus in the piedmont zone of an inselberg in southern India, and other examples have been observed by the writer in southern Africa.

The suggestion is that those compartments that are riddled with fractures are more rapidly and intensely weathered because water can more readily penetrate the mass. They are reduced to plains underlain either by grus or by corestones

Fig. 5.9. Sketch of Blackingstone Rock, eastern Dartmoor, from the northwest with (foreground) quarry in which are exposed joint blocks much smaller than those seen in the hill (after Jones, 1859).

Fig. 5.10. This reservoir is one of several excavations made in the scarp foot zone of Ucontitchie Hill, northwestern Eyre Peninsula, S.A., in the interests of water conservation. The weathered granite, consisting of corestones set in clay, contrasts sharply with the massive unweathered granite of the bornhardt exposed only a few metres away.

set in grus. The massive compartments, on the other hand, remain essentially fresh and are resistant to erosion. Once established, this topography persists and is emphasised, because the residuals shed run-off and tend to remain dry, whereas the plains receive water and, thus, the rocks beneath them become more and more weathered: an example of a reinforcement or positive feedback effect

134

(Twidale *et al.*, 1974).

Variations in fracture density can be explained in various ways. According to Lamego (1938), for example, the distribution of fractures is directly related to the distribution and sense of stress and strain in folded crystalline sequences, with some zones in compression, others in tension. The deeper zones of anticlines or antiforms are in compression, the shallower ones in tension (Fig. 5.11); the converse applies in synclinal structures. Similar stress variations are associated with offset or *en echelon* transcurrent faults (Fig. 5.12) as suggested by such writers as Crowell (1974, 1976). Again, recurrent dislocation of orthogonal sets of regional magnitude could result in distortion of the cubic or rectangular blocks with a tendency to stretching along axes aligned at roughly 45° to the direction of stress (Fig. 5.13; see Twidale, 1980b). Compression and tension

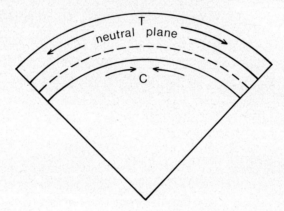

Fig. 5.11. Distribution of stress in an anticline. C - compression, T - tension. (After Price, 1966, p.149).

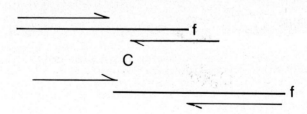

Fig. 5.12. Compression (C) developed in relation to offset (*en echelon*) transcurrent or wrench faults.

do not cancel out but are additive, and are of the same order of magnitude as the stresses applied. Continued dislocation causes fracture propagation in the zones adjacent to the primary fractures (Hoek and Bieniawski, 1965; Hoek, 1968) so that each major block comes to consist of a stressed core set in a fractured zone (Fig. 5.13).

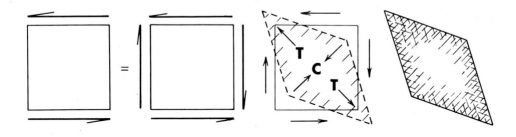

Fig. 5.13. Distortion of cube (with tension, T, and compression, C) and propagation of marginal fractures as a result of shearing.

C. SUBSURFACE INITIATION

Here another element enters the argument, for, whereas Mennell and several others implied fracture-controlled differential weathering at the land surface under epigene attack, most recent workers have alluded to subsurface weathering reaated either to petrological variations or variations in fracture density (e.g. Falconer, 1911; Linton, 1955; Büdel, 1957; Wilhelmy, 1958). In a widely accepted explanation both differential fracture-controlled subsurface weathering and subsequent differential erosion are involved, and for this reason the explanation is frequently referred to as the two stage concept (Fig. 5.14). In these terms inselbergs are compared to boulders (Chapter 4).

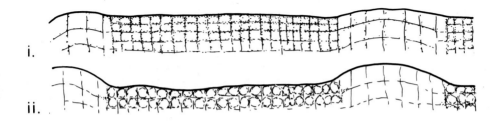

Fig. 5.14. Two stage development of bornhardts (i) differential fracture-controlled subsurface weathering (moisture attack), (ii) differential erosion of the rock mass that has become of contrasted resistance to erosion through the action of weathering.

That subsurface weathering, largely through the agency of water, is involved in the genesis of bornhardts has important implications for the question of the distribution or zonation of the forms. As is made clear below, it has long been accepted that inselbergs in general and bornhardts in particular are climatic forms; the particular climatic connection has varied from time to time and from writer to writer, though savanna or desert environments have found most favour. But groundwaters are ubiquitous, so that in terms of two stage development the first of the requirements necessary for bornhardt evolution obtains over wide areas of the continents, and particularly in stable plains regions.

Moreover, there is increasing evidence that warm, humid conditions, conditions especially conducive to rapid and deep weathering, obtained over most of the continents through much of Mesozoic and Cainozoic time. The strongly differentiated latitudinal zonation of climatic zones characteristic of the present world is both atypical and largely irrelevant to the question of landform development on granitic rocks. Thus in respect of Australia, palaeomagnetic and stratigraphic work (see e.g. Embleton, 1973; Daily et al., 1974) suggests that, though located in high latitudes during the earlier Mesozoic, warm, humid conditions prevailed over much of the continent. The stratigraphic evidence generally favours warm, humid conditions through the Jurassic and Cretaceous (see summaries in Brown et al., 1968) and Kemp (1978) has shown that warm, humid or subhumid climatic conditions obtained even in central Australia through the earlier Cainozoic (see also Twidale and Harris, 1977). Thus optimal conditions for deep weathering have prevailed over wide areas for much of the closer geological past, and the subsurface weathering that took place did so largely unaffected by epigene climatic conditions. Such workers as Linton (1955) and Demek (1964a and b) had good reason to assign the deep differential weathering to which they attributed the initiation of the inselbergs of southwestern England and Bohemia respectively to warm humid conditions of the early Tertiary.

D. ENVIRONMENTS OF DEVELOPMENT

Apart from the controversy as to surface or subsurface initiation, several environments and associated processes have also been invoked either in respect of agencies responsible for exposing or for shaping the landscapes. As their name suggests, it was at one time thought that inselbergs, including bornhardts, were of marine origin. Bornhardt (1900, p. 34) himself considered this view, but brief reflection shows that the argument cannot be sustained in the face of the evidence. It is true that bornhardts are spectacularly developed in coastal and inshore zones in such areas as the Baia Guanabara and on several sectors of the coast of southern Australia (see e.g. Twidale, 1971b), but they are also well represented in interior lands that have not been covered by the sea for scores, even hundreds, of millions of years, and certainly not during the period of inselberg formation. Marine forces have in some areas been responsible for the shaping of inselbergs, for the differential erosion that has produced the inselbergs, and for the asymmetry of some examples (e.g. Pearson Islands - Fig. 1.5), but there is no evidence to suggest that marine forces alone are capable of producing inselbergs in general or bornhardts in particular: it is merely that wave action has exploited weaknesses in the bedrock exposed in the coastal zone, leaving resistant masses in relief.

Explanations of bornhardts based in climate have long been popular. Agassiz (1865) believed that the granitic domes of the Rio area of southeastern Brazil were huge *roches moutonnées* and Le Conte (1873) reached the same conclusion with respect to the domes of the Yosemite in central California. But the 'glacial erratics' of the Rio region that suggested a glacial history to Agassiz are nothing more than corestones, and there is no evidence that the area has been glaciated during the relevant time period. The domes of the Sierra Nevada (Figs. 1.4, 5.4) of southern Greenland (Soen, 1965) and of northern Norway (Kieslinger, 1960) occur in glacial or glaciated regions but, as with marine environments, their climatic context appears to be incidental. Glacial ice is capable of evacuating weathered rock, rather as a bulldozer pushes away unconsolidated debris (Boyé, 1950), and of thus exposing the weathering front, but the end result is basically similar to the landscape exposed by other agencies.

Far more common is the association in people's minds of inselbergs and arid or semiarid lands. In most texts inselbergs are dealt with under the heading of 'The Arid Cycle' or some such title. Passarge (1895) urged that the inselbergs of West Africa were the result of wind erosion of the intervening plains during the Mesozoic, and the aeolian argument was propounded in general terms by such workers as Keyes (1912) and Jutson (1914), though it never found wide favour.

Fluvial erosion of the plains was suggested by Bornhardt (1900) and by a succession of workers from Falconer (1911) to Thiele and Wilson (1915), Bain (1923), and many others. Like Passarge they implied differential erosion but emphasised river work. Bain suggested that alternations of wet and dry seasons were conducive to deep weathering and hence to deep erosion of the plains, and Thomas (1974) and Bloom (1978, p. 324) have both recently asserted that inselbergs are the result of savanna morphogenetic processes and principally fluvial erosion.

That rivers have been responsible for differential erosion of the regolith and for the exposure of the weathering front over wide areas of the continents is undoubtedly correct. In specific instances exposure of the bornhardt masses can reasonably be attributed to wave action, or to glacier ice to nival processes (as on Dartmoor), but rivers have been most widely active.

E. SCARP RETREAT

Riverine erosion of a particular type is invoked by many writers, and particularly by King (1942, 1949a and b, 1953, 1957, 1962), Cotton (1942), Howard (1942), Pugh (1956) and Selby (1977). These workers see inselbergs as *Fernlinge, monadocks de position*, or remnants of circumdenudation remaining after long-continued scarp retreat and pedimentation (Fig. 5.15). Most authors of this persuasion argue that the process is restricted to low lattitude arid and semi-arid lands. Thus, Cotton (1942, p. 90) stated:

> It seems to be a safe assumption that the scarped forms of inselbergs are developed under conditions which do not favour soil formation or forest growth but promote dry weathering and encourage a back wearing process.

Fig. 5.15. Bornhardts as remnants of circumdenudation following long-distance scarp retreat.

Even King (1957), while passionately urging that scarp retreat and pedimentation are dominant wherever running water is dominant (i.e. everywhere save in glacial areas and the dune deserts), nevertheless concedes that the process attains optimal effects in the semiarid tropical and subtropical lands. The distribution of inselbergs seems to support this contention, for they are most commonly and perhaps most typically developed in such areas. The likely reason is that the slope processes active there, and especially the tendency to scarp foot weathering and erosion, leads to the steepening and the recession of slopes and to the formation of a pronounced piedmont angle (see Tricart, 1957; Twidale, 1960, 1967). Furthermore, duricrusts (including gibber) are well preserved in arid and semiarid regions, so that caprocks are more frequently encountered here than in humid lands. Granite is resistant under dry conditions but is very susceptible to alteration when in contact with water: thus in hot, arid and semiarid lands not only is the contrast in erosional vulnerability between high and low topography more pronounced than elsewhere, but the dry granite effectively acts as a caprock so that the development and maintenance of escarpments is theoretically feasible even in granitic terrains. Finally, the rates of geomorphic change are so slow (Corbel, 1959) that the forms are better preserved in arid lands than elsewhere.

For the supporters of this hypothesis the rounded form of bornhardts is due to differential weathering under epigene attack, and the sheet structure is consequent upon the development of the domical shape.

F. EVIDENCE AND ARGUMENT

Several different hypotheses have been advanced in explanation of bornhardts, some residing in structure, others in external processes. What is the evidence?

(i) Fracture-controlled margins

First, as mentioned previously, many bornhardts are defined by prominent vertical or near-vertical fractures (Fig. 5.1). It is surely asking a great deal, and perhaps too much, of coincidence that scarp retreat should have just arrived at and apparently stabilised on these fracture zones. Surely in some instances it would have regressed into non-fractured rock?

(ii) Regional settings

If bornhardts and related forms are in any way due to crustal stress they ought to be disposed in patterns that are geometrically conformable with regional tectonic style. They are. Thus, on northwestern Eyre Peninsula the residuals occur on clear N.W.-S.E. trending ridges that parallel known regional fractures and in detail display clear alignments and fracture patterns related to these broader trends (Figs. 5.16 and 17; also 5.1 (i)). S.A. Tomich (pers. comm., 4 Jan. 1982) reports that in the Yilgarn Block of the southwest of Western Australia granitic and 'greenstone' (basic and ultrabasic volcanics) bornhardts appear to be confined to the vicinity of regional lineaments or regional fracture zones of Archaean age. Again, in the Transvaal the granite outcrops, and hence inselbergs, occur at the intersections of the axes of regional flexures (Anhaeusser, 1973).

(iii) Upland settings

Some bornhardts occur in plains settings, and, as they include some of the most spectacular examples known, they are given greater significance than they perhaps deserve. If bornhardts were the last remnants surviving after scarp retreat and pedimentation have run their course, the residuals ought to be found only in plains contexts, and they ought also to be restricted to major divides. They are not. They occur in valley floors and on valley side slopes as well as on divides (Figs. 5.18 and 19). In reality, many are found in upland settings - in the Valley of a Thousand Hills in Natal, in the Yosemite and Domeland, both in the Sierra Nevada of California, in southern Greenland, in the Rio region, and in the Kamiesberge of Namaqualand. In the Darling Ranges they are apparently being exposed as the duricrust-capped upland is dissected. These residuals surely argue against the suggestion that the forms are inherently *Fernlinge*.

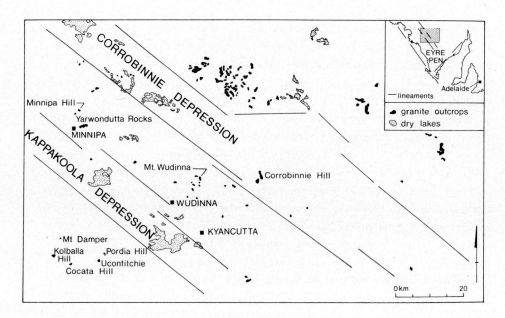

Fig. 5.16. Northwestern Eyre Peninsula, showing major fractures and granite outcrops.

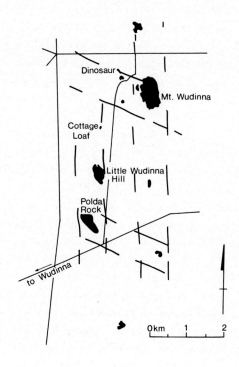

Fig. 5.17. Occurrence of residuals in aligned zones near Mt Wudinna, S.A.

Fig. 5.18. The Witrivier valley of central Namaqualand.

Fig. 5.19. Field sketch of part of the Swakoprivier valley, central Namibia,
showing granitic and other inselbergs at various levels in the landscape.
X - palaeosurface.

(iv) Deep weathering and contrast between hill and plain

As noted earlier (Chapter 3) deep weathering of granite beneath plains has
been reported from several parts of the world. On the other hand, there is a
strong contrast between the weathering of the bedrock beneath the plains and
that of the bornhardts. Thus, in the excavation already referred to (Fig. 5.10)
at the margins of Ucontitchie Hill, the granite has suffered marked differential
fracture-controlled weathering, with corestones set in a matrix of weathered
rock. By contrast, only a few metres away, the massive rock exposed in the
bornhardt remains essentially fresh. Similar contrasts between fresh rock in
uplands and weathered rock beneath the plains have been observed at several
other sites (see earlier, p. 132).

(v)   Incipient domes

If bornhardts are, indeed, initiated in the subsurface by differential
weathering, as is implicit in the two stage hypothesis, there ought to be examples
of all stages of their formation, including some domical masses formed of intrinsically fresh rock, and that have either just been exposed or are present just
beneath the natural land surface.

Several examples have been noted in the literature and in the landscape.
From near Ebaka, in South Cameroon, Boyé and Fritsch (1973) describe a quarry
opened to provide ballast for a railway line. The excavation exposed a domical
mass of fresh granite surrounded by weathered rock (Fig. 5.20). Its crest was
situation 8-10 m beneath the land surface and the surface of the fresh mass
dipped away in all directions. This is surely an incipient bornhardt awaiting
natural exposure but revealed by artificial excavations. The dome has not been
shaped by epigene processes and then buried, for the cover material is weathered
granite *in situ*.

Fig. 5.20. This dome was exposed as a result of excavations at Ebaka, South
Cameroon (M. Boyé).

Near the Leeukop, south of Potchefstroom in South Africa, the crest of a dome
is exposed in a shallow depression excavated as a water storage (Fig. 5.21). In
the Vredefort brick quarry, in the northeastern Orange Free State, the weathered
granite has been excavated, revealing mainly corestones but, in one corner of
the quarry, part of an incipient dome. At Halfway, between Johannesburg and

Pretoria, what was a small rock platform has been revealed by road excavations
to be the crest of a small dome (Fig. 5.22). Similar features have been noted
at Buccleuch, south of Johannesburg; between Nuwe Smitsdorp and Pietersburg in
the northern Transvaal; at several sites between Vanrhynsdorp and Nuwerus in
Namaqualand; at the Pomona Quarry, near Salisbury, Zimbabwe; and at several
sites in the southwest of Western Australia, particularly in the Darling Ranges,
where forms like Sullivan Rock have only just been exposed.

(vi) Subsurface initiation of minor forms

If minor forms are initiated beneath the land surface, it follows that the
host mass must also have evolved beneath the land surface. As is made apparent
in several later chapters, there is irrefutable evidence that some minor features
characteristic of granite domes have been initiated at the weathering front.
Thus excavations at the margins of several inselbergs on northwestern Eyre
Peninsula (Fig. 5.23) show that both basins and gutters are developed at the
weathering front (Twidale, 1971a, p. 90; 1976b, pp. 203-204; Twidale and Bourne,
1975a). There is no possibility of their having formed subaerially and then
suffered burial, for the overlying material is grus *in situ*. Some platforms,
for example that which borders Corrobinnie Hill (Fig. 1.26), are in detail
irregular due to the development of numerous shallow basins or saucer-shaped
hollows and gutters. Some patches of grus remain and there are also small
boulders that can be interpreted as erstwhile corestones released through the
erosion of the grus.

Fig. 5.21. The Leeukop is a gneissic dome located near Potchefstroom in the
western Transvaal, R.S.A. Nearby there are several low domes and what appears
to be the crest of one (X) exposed in a shallow excavation beneath the plains
that surround the residual hills.

Fig. 5.22. At Halfway, between Pretoria and Johannesburg, a small incipient dome has been further exposed as a result of engineering works in connection with the freeway.

Fig. 5.23. The flutings and runnels that score Dumonte Rock, on northwestern Eyre Peninsula, clearly continue beneath the natural land surface which is still well-defined (X-X) here in an excavation made to create a water storage.

Similarly, in their account of the Ebaka Quarry, Boyé and Fritsch (1973) record that the newly exposed domical surface is scored by gutters and basins.

(vii) Flared slopes and stepped inselbergs

Many of the inselbergs of southern Australia display steepened, or flared, basal slopes (Fig.1.23). There is convincing evidence that flared slopes are a particular

form of the weathering front, and that they are initiated beneath the natural
land surface (see Chapter 9), for at Yarwondutta Rock, at Chilpuddie Hill, both
on northwestern Eyre Peninsula, and elsewhere, such concavities can be seen in
excavations beneath a natural cover of grus *in situ*.

Where such flared slopes are exposed, as they are at many sites, it is clear
that the upper shoulder of the flare marks the location of a former hill-plain
junction. On some inselbergs, including such bornhardts as Mt Wudinna and
Ucontitchie Hill on Eyre Peninsula, and Kokerbin Hill and Hyden Rock on the southwest of Western Australia, linear subhorizontal zones of flared slopes and
associated platforms, steepened slopes and tafoni occur at various levels, so
that the residuals have a stepped appearance (Twidale and Bourne, 1975b;
Twidale, 1982).

Yarwondutta Rock provides a clear, compact example (Fig. 5.24). Former hill-
plain junctions are indicated by the upper shoulders of the two flared slopes
preserved on the northern slope of the Rock, and in these terms:-
(a) Yarwondutta Rock has emerged and grown as a positive relief feature as a
result of the episodic or phased lowering by streams of the surrounding plains.

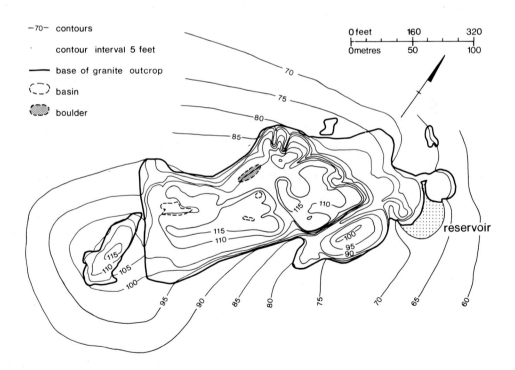

Fig. 5.24. Contour plan of Yarwondutta Rock, showing stepped northern slope and
fracture-controlled outlines and clefts.

(b) This emergence has taken place not gradually but in distinct phases or epi-
sodes, for the flared weathering fronts imply comparative standstill, and time
in which scarp foot weathering has taken place, while their exposure indicates
stream rejuvenation and landscape revival.

(c) The process of scarp foot weathering and the first stage in the development
of another set of flared forms is again in process for they are encountered in
excavations (see Chapter 9).

(d) Remnants of landforms related to earlier phases or cycles of development
have clearly persisted. Yarwondutta Rock appears at one time to have been merely
a low platform with, at most, a few boulders or sheet remnants standing above the
general level, but that upper surface, as it now is, has survived at least two
subsequent phases of weathering and erosion.

(e) There has been some backwearing of the bounding slopes, not by river action
- the residual is too small to generate much run-off - but by weathering in the
scarp foot zone. It amounts to only a few metres (see also Twidale, 1978b), and
is of quite a different order to the scores, even hundreds, of kilometres demand-
ed by the scarp retreat hypothesis and by inselberg landscapes in many parts of
the world.

(f) The residual has grown episodically in relief amplitude in time. This
explanation meets the objection to the two stage hypothesis, and due to Bakker
(1958) and King (1966), that the maximum depth of weathering recorded in a given
region is commonly much less than the height above the plains of the highest
inselbergs in that region. The implication was that some inselbergs are too
high to have been initiated in the subsurface, but the workers mentioned were
thinking in terms of a single cycle, not of multicyclic or multiphase development.

(viii) Age of inselbergs

If inselbergs in general, and bornhardts in particular, are the last remnants
surviving after long-continued erosion (scarp retreat) no bornhardt ought to be
older than the duration of a geomorphic cycle, or the time taken to reduce a
land mass to baselevel. Estimates of the duration of a cycle vary widely (see
Gilluly, 1955; Linton, 1957; Schumm, 1963; also Twidale, 1976a) but there is
general agreement that even after due allowance is made for isostatic recovery
large areas of high land would be reduced to baselevel in periods of the order
of 35-40 Ma. Thus no inselberg ought be of greater age than this; in stratigraphic terms no inselberg ought to predate the late Eocene and most ought to
be much younger.

Disregarding those several forms that are of exhumed character (see e.g. du
Toit, 1939; Twidale, 1968a, pp. 121-123, 1981b; Williams, 1969), several
writers, including Willis (1934) and Dresch (in Birot, 1958), have suggested
that some inselbergs are of considerable antiquity. Bornhardt massifs like the

Everard Range stand above plains on which silcrete developed and which are therefore (Wopfner and Twidale, 1967; Wopfner *et al.*, 1974) at least of early Cainozoic age. Twidale and Bourne (1975b) have suggested that the crests of some of the bornhardts of northwestern and northern Eyre Peninsula may be of Mesozoic age, and Twidale (1982) has adduced evidence to show that some of those of Western Australia are of late Mesozoic or earliest Tertiary age (Fig. 5.25).

This is not to suggest that all inselbergs are old forms. On the contrary, many are clearly youthful, having been only recently exposed. But others are ancient, even in geological terms, and they stand as mute witnesses against scarp retreat. On the other hand, they are compatible with the two stage hypothesis, though their survival poses problems (Twidale, 1976a).

(ix) Occurrence in multicyclic landscapes

Bornhardts characteristically occur in multicyclic landscapes and can be seen in process of exposure from beneath a lateritised land surface of late Mesozoic - early Tertiary age (Fairbridge and Finkl, 1978) in the Darling Ranges of Western Australia; in the Valley of a Thousand Hills in Natal, and in the eastern Transvaal (Fig. 5.3), where they occur beneath the (early Tertiary) land surface; and so on. Jessen (1936) clearly related bornhardts and planation surfaces in Angola, and Twidale and Bourne (1975b) have correlated inselbergs and duricrusted surfaces on Eyre Peninsula, and remnants of old planation surfaces remain high in the bornhardt landscapes of southern Greenland (Soen, 1965 - see Figure 5.26).

The link between bornhardts and multicyclic landscapes is twofold. First deep differential subsurface weathering arguably requires a period or periods

Fig. 5.25. The Humps is a complex granite bornhardt in the southwest of Western Australia. The surrounding plains are underlain by laterite which (foreground) has been quarried for use as a road metal. Clearly then the upland is at least as old as the laterite, i.e. late Mesozoic or earliest Tertiary.

of standstill and baselevelling, such as is evidenced by palaeosurfaces.
Second, the erosion achieved during x + 1 cycles is consistent with the suggested
exposure of deeper compressional zones of the antiformal structures (Fig. 5.11)
that are bornhardts with well developed sheet structures. Large radius domes
with few visible fractures and few residual boulders or slabs, domes like Little
Wudinna and Polda on northwestern Eyre Peninsula, and the Lightburn Rocks and
Childara Rock in the eastern Great Victoria Desert, are interpreted as domes
located below the neutral planes of antiforms. In residuals like Ucontitchie
Hill, on the other hand, with several layers of sheet structure and exposed
boulders (see Fig. 2.27 (a)), the neutral plane may be coincident with the surface of the main mass but lie below the various sheets represented only by large
blocks and boulders.

The two stage mechanism is consistent with the known distribution, global,
regional and local, of bornhardts, for if bornhardts are structural forms, as is
implied, they can be expected to occur wherever structural conditions are suitable, regardless of past or present climate, to be exposed in a range of topographic situations, and so on. For many workers bornhardts are simply gigantic
boulders that have developed by differential subsurface weathering and subsequent erosion. Falconer (1911, p. 246) long ago provided an admirably succinct

Fig. 5.26. Part of Sermasoq, southern Greenland, showing remnants of old
planation surface preserved in granitic rocks (Oen Ing Soen).

summary of the two stage concept:

> A plane surface of granite and geniss subjected to long-continued
> weathering at base level would be decomposed to unequal depths, mainly
> according to the composition and texture of the various rocks. When
> elevation and erosion ensued, the weathered crust would be removed, and
> an irregular surface would be produced from which the more resistant
> rocks would project. Those rocks which had offered the greatest resis-
> tance to chemical weathering beneath the surface would upon exposure
> naturally assume that configuration of surface which afforded the least
> scope for the activity of the agents of denudation. In this way would
> arise the characteristic domes and turtlebacks which suffer further
> denudation only through insolation and exfoliation. Their general ellip-
> tical outlines, which Merrill would ascribe very largely to the influence
> of crustal stress and strain, are probably in great part due simply to
> the modification by weathering of original phacolitic intrusions.
> (Falconer, 1911, p. 246).

Bornhardts are domes in both the structural and topographic sense. They are
Hartlinge, monadnock de dureté, or *de résistance*. Because of their structural
origin, their subsurface initiation, and the ubiquitous occurrence of groundwater to which they owe their primary causation, bornhardts are essentially
azonal forms. This is not to suggest that climatic factors have no effect. The
residuals of southern Greenland, for example, are more angular than most, no
doubt reflecting nival and glacial action. Again, in the humid tropics the
stabilising effect of vegetation allows distinct debris slopes to form and persist, in contrast with the well-defined piedmont angle, the abrupt transition
from hill to plain, developed in arid and semiarid lands. In part this is due
to preservation under conditions of slow change in the aridity of recent times,
in part to the excellent development of the piedmont angle in arid and savanna
lands (Twidale, 1967).

G. DOMICAL FORM

Regardless of which interpretation of bornhardt origin is accepted, the
reason or reasons for the domical shape of the hills has given rise to further
controversy.

(i) Weathering

One school of thought suggests that a spherical or hemispherical shape is an
equilibrium form for a residual mass. Certainly a sphere contains the greatest
volume per unit of surface area and, from this point of view, a hemisphere
presents the minimal surface area for a projecting mass exposed to weathering
and erosion. This explanation was suggested by Mennell (1904) who pointed out
that outstanding edges of fracture-defined blocks are readily removed; Falconer
(1911, p. 246) thought that the inselbergs of Nigeria 'naturally assume that
configuration of surface which afforded the least scope for the activity of the
agents of denudation', and White (1945) considered that the domes of the southeastern Piedmont of the U.S.A. are the product of granular disintegration and

rounding by weathering. Penck (1924; 1953, pp. 141-143) also argued, albeit in
a general context, that rounding represents a stabilising tendency.

(ii) Sheet structure (See also Chapter 2)

Bornhardts are invariably associated with sheet structure but whether these
arcuate fractures give rise to the domed shape or whether they are induced by
it is debatable. Two diametrically opposed views of the relationship between
the form of the land surface and the geometry of sheeting joints have evolved
over the past 130 years or so.

Some interpret the joints and associated sheet structure as a primary feature
of the rock which has closely determined the gross morphology of the land surface. According to this view the joints were first developed in the bedrock and
the shape of the land surface is a response to this internal structure. As
Merrill (1898, p. 245) put it:

> ... with many geologists these joints, in themselves, would be accepted as
> due to atmospheric action. In the writer's opinion they are, however, the
> result of torsional stress and once existing are lines of weakness which
> become more and more pronounced as weathering progresses.

According to Merrill (1898, p. 245), the boss or dome-like form of the bornhardts is 'incidental and consequent' on internal structure. The earliest pro-
ponent of this general endogenetic interpretation was de la Beche (1839, p. 163),
but several others besides Merrill, also subscribed to the theory last century.

Nowadays, however, such endogenetic theories of sheet structure development
find little support. There have always been those who took a radically different stance and saw the geometry of the sheeting joints as a response to the form
of the land surface, and this view now prevails, though it is not necessarily
correct on that account.

In detail, many mechanisms of and explanations for sheet structure have been
proposed, but all fall into one of these two major categories - exogenetic or
endogenetic.

(a) Exogenetic theories of origin

Insolation: As rocks are poor conductors of heat it has been argued that
solar radiation heats the outer exposed zones of rock which expand and become
detached from the main mass, forming more-or-less thick slabs or sheets. Shaler
(1869) propounded this view in explanation of sheet structure he had observed
in New England (U.S.A.), as did MacMahon (1893) with respect to the pseudobedding of Dartmoor, and Tyrrell (1928) in connection with sheet structure
developed in granite on Arran, Scotland. But because the effect of the sun's
radiation penetrates only a few centimetres at most into the rock, whereas
sheet jointing extends to considerable depths, this view may safely be
discounted.

Chemical weathering: The gradual infiltration and penetration of meteoric waters into rocks near the land surface has frequently been called upon in explanation of the flaking and spalling of rock masses (see e.g. Blackwelder, 1925, 1933). Where the chemical alteration of rocks results in increased volume and, hence, pressure, this appears feasible. However, as has been pointed out elsewhere (Ruxton, 1958; Trendall, 1962), not all chemical alteration leads to volume increase and, hence, to expansive pressure and rupture. Also, if the weathering were held to precede and to give rise to the fracturing, it is legitimate to ask why chemical attack is concentrated upon and restricted to a few gently arcuate planes.

Many of the massive slabs and wedges of rock involved in sheet structure display no sign of chemical alteration. Some do, but such mineral alteration can more readily be explained in terms of weather induced by moisture attack subsequent to fracture development.

Offloading or pressure release: Most workers agree with Chapman (1956) that all joints and, indeed, all rock fractures are an expression of erosional offloading in the sense that at depth other stresses are subordinate to the vertical pressure exerted by the superincumbent load, and that it is only through the release of vertical pressure that the other stresses are manifested as obvious fractures. But a basically different interpretation of sheeting joints which attributes them solely and wholly to pressure release without the previous application of stress, has long found favour.

The gist of the pressure release, or erosional offloading, hypothesis is that rocks which cool and solidify deep in the earth's crust (as, for instance, granites, whether of metasomatic or igneous origin) do so under conditions of high lithostatic pressure. That there are widespread granite outcrops is itself proof of deep erosion, for it is generally agreed that however it is formed granite originated at considerable depths in the crust. Decompression achieved through the removal of superincumbent load is said to cause the development of radial stress which is relieved by the development of fractures tangential to the land surface; these are the sheeting joints of many outcrops. The fundamental premise of the hypothesis is that the form of the land surface in broad terms determines the geometry of the sheet jointing, for it is in relation to this that the radial stress develops.

First outlined by Gilbert (1904), the offloading hypothesis has many adherents, of whom Matthes (1930), King (1949a), Waters (1954), Hurault (1963), Ollier (1965), Hack (1966) and Bateman and Wahrhaftig (1966) may be mentioned. Soen (1965, p. 12) points to the crux of the offloading argument when he states that 'From a genetic point of view sheeting should ... be distinguished from primary and tectonically imposed joints in the granite'.

Sheeting involves the splitting or subdivision of the granite mass into more-or-less thick slabs oriented parallel to the land surface; thus sheeting is in these terms a secondary feature formed after the development of the topography. Kieslinger (1960, p. 273), for example, relates the geometry of sheeting on the Bohemian Massif to a hilly topography of Tertiary age.

Thomas (1965) suggested that sheet structure developed beneath the land surface, before the host residual was exposed, though the same author subsequently argued (Thomas, 1967) that sheet structure, like other minor landforms, is due to disintegration under epigene attack. In either case, however, sheet structure was seen as due to offloading.

The general parallelism of sheeting joints and land surface can be taken as lending support to the offloading hypothesis, though the interpretation can, of course, be reversed with equally satisfactory logic. Also, there is considerable evidence that offloading can permit and induce the formation of relatively thin slabs or sheets of rock close to the land surface. Sheets of rock have been developed in cirque headwalls and in the floors of recently deglaciated valleys (see Lewis, 1954; Linton, 1963; Gage, 1966; Harland, 1957) and fractures formed in relation to older glacial and more recent riverine valleys, as for instance in northern Italy (Kiersch, 1964). In anisotropic rocks, local unloading or unbuttressing has caused the development of features morphologically similar to sheet structure. For instance, in the Torrens Gorge, east of Adelaide, steeply dipping gneisses are seen to bulge outwards in moderately thick slabs, some of which are partially detached from their neighbours (Stapledon, 1966). This is, however, a local effect and is due to the removal, partly as a result of natural erosion, partly as a result of quarrying, of bedrock and the partial failure of the newly exposed gneisses under their own weight.

Though it is logically persuasive the pressure release hypothesis in the sense outlined by Gilbert and adopted by many later workers, namely, that offloading is the sole cause of sheet jointing, may be called to question on several grounds.

First, simple triaxial tests show that compression and decompression of essentially isotropic materials does not cause fracturing save in special circumstances which are unlikely to be found in nature; such fractures are unlikely to develop in the context of slow erosional unloading. Even in anistropic materials it appears that several cycles of compression and decompression can be applied to unconfined specimens without fracturing before, with increased loads, the material ruptures in fatigue (Ingles et al., 1972). Wolters (1969, p. 61) has also questioned why unloading should cause fracturing, and Brunner and Scheidegger (1973) also reject the suggestion after mathematical analysis.

Second, it is difficult to understand why, if expansive stress developed during erosion, it has not been accommodated along pre-existing lines of weakness. Both Jahns (1943, p. 75) and White (1946) point out that sheeting is absent from well-jointed granites and 'most geologists accept this fact as evidence that the force of expansion in the rock has been dissipated by slight movements along the joint planes' (White, 1946, p. 5). But evidence from many areas shows that orthogonal joints predate sheet structure and, in addition, there are other potential slippage planes in rocks. Strain could also be taken up by grain boundary sliding, as described by Gifkins (1959, 1965), and along crystal cleavages for example. Thus there is no reason why any later stresses could not have been accommodated along such slippage planes without the development of new fractures.

Third, sheet jointing occurs in rocks which show no signs of ever having been under great stress. In the Colorado Plateau sheet structure has developed in sandstone which has been only gently warped (Bradley, 1963). These and other examples indicate that sheet structure has formed on strata which have not suffered deep burial and high lithostatic pressures.

Fourth, the association of sheet structure with inselbergs of the bornhardt type is irrational if the former are interpreted as a consequence of offloading without the application of compressional stress. Sheet structure is supposed to be a manifestation of radial expansion, whereas the field evidence, both in gross and in detail, shows that the inselbergs are masses in compression. This is indicated by the condition of the joints within the inselberg masses, which are tight, and commonly take the form of discontinuous hairline cracks.

The steep dip of sheeting joints in the vicinity of vertical or near-vertical fractures (Figs. 2.29 and 3.28) suggests that they are developed in relation to large orthogonal blocks that have been horizontally compressed, which is inconsistent with the suggestion that the residual mass is in or has been in a state of tension (radial expansion).

If the inselbergs were indeed developed on masses of granite that were decompressed and relaxed, their joints would be open and the rock masses would not survive weathering and erosion and would not, therefore, stand out as residuals. This line of argument is tacitly sustained by the observations of Hack (1966) who, working on large-scale crystalline circular structures in the central Appalachians, noted that the expansive stresses to which he attributed the sheet jointing developed in rocks had also caused earlier (tectonic) joints to gape, thus allowing the weathering and erosion of these circular masses; these do not however stand out as bornhardts, merely as concentric fracture patterns in hilly terrain.

Fifth, although it is conceivable that the parallelism between sheet jointing and land surface need not be perfect, it is difficult in terms of the offloading hypothesis to explain inverse relationships. Similar features have been observed on Dartmoor and in Yosemite National Park in the vicinity of Tenaya Lake (Twidale, 1971a, p.66 - see Fig. 2.30).

Sixth, there are many local and detailed items of evidence which argue against the offloading hypothesis. For instance, there is an inconsistency between the age of erosional features said to be the cause of sheeting joints and the inferred age of the joints on Dartmoor (Twidale, 1971a, pp. 67-68): the former are geologically youthful whereas the latter are of considerable antiquity, so that it seems at least as reasonable to suggest that the joints determined the form of the land surface as the converse (see also Gerrard, 1974).

Thus there are many lines of argument and many types of evidence which, together, strongly suggest that the offloading hypothesis should not be uncritically accepted as an explanation of sheet jointing. Of these several considerations, undoubtedly the most significant from a geomorphological point of view is that inselbergs are rock masses in compression, whereas radial expansion and tensional stress are implied by the offloading hypothesis.

(b) Endogenetic theories of origin

Plutonic injection: Several writers, including some of the earliest to consider the problems of sheet jointing (de la Beche, 1839; Whitney, 1865; Vogt, 1875; Harris, 1888, p. 106), related sheet structure to the stresses imposed on magmas during injection or emplacement and, hence, to the shape of the original pluton. De la Beche (1839, p. 163) noted the parallelism of sheeting joints and the margins of the igneous body on Dartmoor, and cited observations by Boase (1834) to this effect. Brammell (1926) attributed sheet structure to a combination of stresses developed during emplacement of the granite mass and later cooling.

Although it may apply at a few sites, this suggestion cannot stand as a general hypothesis, for inselbergs and associated sheet jointing are well developed in sedimentary rocks.

Metasomatic expansion: Jones, in 1859, suggested that the nodular form or concentric structure of many crystalline masses was responsible for the dome-like shape of Blackingstone Rock, an inselberg (or tor) on eastern Dartmoor. Brajnikov (1953) also envisaged that the domed inselbergs or *morros* of southeastern Brazil are gigantic floaters of solid rock developed as foci of compression consequent on volume changes developed during the processes of metamorphism. However, and as stated earlier, there is no reason why decompression should cause fracturing. Also, sheet jointing occurs in rocks which clearly have not suffered metamorphism.

Vertical uplift: Because many granite masses are areas of distinct negative gravity anomaly (Bott, 1953, 1956; Rowan, 1968), it has been suggested that these masses tend to rise as diapirs through the superincumbent rocks to form gneiss domes.

Gneiss domes are structures developed in migmatised rocks, that is, in granitic rocks consisting partly of igneous, partly of metamorphic, materials.

Foliation is well developed and forms concentric patterns with quaquaversal dips. Gneiss domes have been described from many areas (see Chapter 2), and prominent circular structures in gneissic rocks occur in such areas as Zimbabwe (MacGregor, 1951), North Carolina (Hack, 1966) and French Guyana (Choubert, 1974).

Some workers interpret gneiss domes as due to repeated injections of magma during separate orogenies or distinct phases of the same orogeny (Eskola, 1949). Others, and notably Kranck (1957), consider that the structures can be explained in terms of a single phase of compression resulting in upward migration of migmatic material to form a dome which then spreads laterally to form a mushroom-shaped mass.

Whatever interpretation is placed on the structures in detail, vertical upthrust of granitic material is involved. Ollier and Pain (1980) claim that the dome that is Goodenough Island, off the northeastern coast of Papua New Guinea, and Dayman Dome, on the adjacent mainland, are still rising. In these structures, sheet jointing is construed as due to vertical movement and the development of radial stress, rather than to lateral compression; sheet jointing is caused by the radial stretching introduced during uplift. A number of objections can be levelled against the hypothesis. For example, sheet structure occurs in sedimentary and volcanic rocks which have not been subjected to doming, and, if the sheeting joints are stretching planes, it is difficult to explain the preservation of inselbergs as well as the other field evidence indicative of compression.

Lateral compression: Faults and sheet jointing commonly occur in association in the field, for instance, in southwestern England (see Blyth, 1957, 1962; Dearman, 1963, 1964), southeastern Brazil (Lamego, 1938; Barbier, 1957; Birot, 1958), northeastern U.S.A. (see Twidale, 1971a, pp. 73-76, and Fig. 4.16 (a)) and in the western Gawler Ranges.

Some faults are a manifestation of compression in the crust, so that there is a rational link between compression and sheet structure. H. Cloos (1929) pointed out that shearing is associated with the margins of many intrusive masses and that curvilinear fractures or secondary shears develop in association with the primary fractures (see also Balk, 1937, p. 101 *et seq.*), while E. Cloos (1955) has provided experimental data supporting this thesis.

It has also been suggested (Dale, 1923; Twidale, 1964, 1971a, pp. 69-77) that sheeting joints may be an expression of lateral compression which results not, or not only, in faulting, but rather in strains which, after the erosional removal of superincumbent load, cause the development of arched fractures or joints of arcuate geometry - as 'a series of undulating fractures extending entirely across' (Dale, 1923, p. 35) a rock mass. Lamego (1938) clearly envisaged a similar situation obtaining in the Rio de Janeiro region (Fig. 5.8; see

also Fig. 3.13 (a)). It is of interest that in 1904 G.K. Gilbert, the author
of the pressure release hypothesis, suggested that Stone Mountain in Georgia,
U.S.A. was due to compressive strains (see Dale, 1923, p. 29) and that several
other geologists who examined granitic domes in the northeastern United States
concluded that the granite was in compression (Niles, 1872; Emerson, 1898,
pp. 63-65). Similar conclusions were reached with respect to a granite exposed
in a quarry at Quenast in Belgium (Hankar-Urban, 1906).

In some areas, e.g., New England, U.S.A., there is evidence of double sheet
structure (Dale, 1923, p. 36), that is, two sets of sheet joints, the strikes of
which intersect. This, surely, is explicable only in terms of two phases of
compression applied from different directions.

As earlier mentioned many sheets have developed triangular wedges on their
exposed faces. Thin slabs of rock, some of which have also been displaced, are
associated with several of them. The wedges of rock surely demonstrate lateral
dislocation. There are slickensides, indicative of differential movement along
sheeting planes in some areas, for instance, in the Yosemite. The wedges are
interpreted as having been formed and displaced as a result of compressive stress,
and the resultant development of arcuate fractures and differential movement
along them (Fig. 2.29).

It has been shown by *in situ* stress measurements that the Australian continent
is in a state of substantial horizontal compression (Denham *et al.*, 1979), and
many other examples of pronounced compressive stress in the horizontal plane,
stress far greater than suggested by theoretical considerations (Talobré, 1957),
have been recorded. The excessive stress may be attributed to relict compression
derived from past orogenies, though the role of continuing or modern earth movement of compressive type should not be overlooked; there is, indeed, much and
increasing evidence of contemporary compression in the crust (see, for instance,
Gilluly, 1949; Bendefy, 1959).

Isaacson (1957) reported that at a depth of 1056 m in one of the shafts of
the Kolar goldfield the theoretical stresses ought to be 313.538 kg/cm^2
vertically and 134.976 kg/cm^2 horizontally, whereas, in reality, the measured
stresses were 409.146 kg/cm^2 and 471.01 kg/cm^2 respectively. Expansion consequent on the release of inherent stress caused a shaft 3.81 m diameter at a depth
of some 3048 m to decrease by 0.5334 cm in a north-south direction and by 1.16 cm
east-west. Moye (1958) obtained similar results in the Snowy Mountains of N.S.W.

Coates (1964) reported deformation in a tunnel some 90 m below the surface in
southern Ontario. There was rapid expansion of the walls (contraction of the
tunnel) in the first forty days after excavation and the 160-190 days after
exposure, so that, after 240 days, there had been up to 4.6 cm of lateral
expansion. Vertical movements were noted, but they were consistently small.

Also, the evidence derived from gneissic (as opposed to gneiss) domes -

domes developed in gneissic rocks - is suggestive. In the Reynolds Range, central Australia, and in the Kamiesberge of central Namaqualand, R.S.A., for example, bornhardts are well developed on gneisses (Fig. 5.27). Sheet structure is well represented and the domes are morphologically indistinguishable from those formed on granite. The gneissic character of the rocks does not become apparent until the sheet structure breaks up, when large foliation plates are formed (Fig. 4.11). This is compatible with the sheet structure being due to internal stresses, but in terms of offloading and tensional stress, foliation plates would surely predominate and be developed to the exclusion of other partings.

Thus, though any of several possible explanations of sheet jointing may be valid in particular areas, the hypothesis offering the best general explanation is that involving lateral compression, induced by horizontal stresses, either relic or modern, and the manifestation of stress patterns near the surface when vertical loading is decreased by erosion. Such an explanation accounts for many details of the field evidence, is consistent with measured stress conditions, and offers a comprehensive view of the preservation of inselbergs and the sheet structure widely associated with them.

Alternatively Brunner and Scheidegger (1973) argue that compressive stress induces tensional cracks developed at the boundaries of Griffith fractures, and claim that the pattern of tension fractures resulting from such stresses can explain such features as their parallelism with the surface, their increase in

Fig. 5.27. Bornhardts in gneiss, Reynolds Range, N.T.

thickness with depth, and their disappearance in depth. This is an interesting suggestion, worthy of testing in the field.

H. OTHER INSELBERGS

In the previous section it is argued that, though they are in some degree convergent forms that originate in various ways, many, perhaps most, bornhardts develop in two stages, the first involving differential fracture-controlled subsurface chemical alteration, the second, the erosional exploitation of the weathered compartments so formed. What, then, of the block-strewn nubbins and the castellated forms or castle koppies? How do they develop, and what, if any, is their relationship with the domed forms? Do they reflect structural varia- tions, or are they a manifestation of different, climatically controlled, processes?

Climate does not offer a complete explanation, for, although castle koppies are well represented in such cold or seasonally cold regions as Dartmoor, the Massif Central and the high plains of the Pyrenees of France and Andorra, and of the Colorado Rockies, domes occur in similar climates. For instance, there are domes flanking the valley of the Thompson River, in Colorado (Fig. 5.28 (a)), only a few score kilometres distant from the castellated forms shown in Fig. 5.28 (b). Koppies also occur in such warm, subhumid regions as eastern Zimbabwe and central Australia (e.g. Devil's Marbles - Twidale, 1980c) and in arid areas like the Hoggar Mountains of north Africa (Fig. 5.29). Nubbins are the most common type of granitic inselberg in the seasonally humid tropical regions of northern Australia, but are found also in semiarid parts of the eastern Mount Lofty Ranges, in the arid Pilbara of Western Australia (see Fig. 5.30), and in central Namaqualand, R.S.A., as well as in a coastal context in several parts of southern Australia.

Both castellated and block-strewn forms are better represented in some climates than in others, but in neither case can it be argued that they are restricted to a particular zone. Furthermore, the two types are commonly coexistent, koppies and domes occurring side-by-side in central and northeastern Zimbabwe and in the Colorado Rockies, and nubbins and domes in the western Pilbara, to cite but a few examples amongst many. Again, elements of the two forms are developed in the same hill at some sites. Thus, in the southern face of Blackingstone Rock, on eastern Dartmoor, huge quadrangular blocks are exposed, and the form is castellated, but the northern side is dominated by massive convex-upward slabs of rock that give a domical shape to the hill (Fig. 5.31). According to Holmes (1918), Mt Kobe, in Mozambique, is similarly two-faced. Again, some of the residuals of the Devil's Marbles complex of central Australia consist of angular, castellated residuals surmounting large radius domes (Fig. 5.32). Similar features are developed in the Albany district of southwestern

Fig. 5.28 (a). Above are seen domes exposed in valley side walls of Thompson River valley, near Estes Park, Colorado, U.S.A. Below (b), castle koppie standing above highlevel planation surface, Rocky Mountain National Park, Colorado.

Fig. 5.29. Castellated inselberg, Hoggar Mts, southern Algeria (P. Rognon).

Fig. 5.30. Nubbin in granitic gneiss, western Pilbara, W.A.

Fig. 5.31. Blackingstone Rock, eastern Dartmoor, seen (i) from the northwest
(ii) from the south.

Western Australia, at Remarkable Rocks, Kangaroo Island (Fig. 5.33), and in the
Shashe district of central Zimbabwe. At Waulkinna Hill and in the Mt Bundey
district, near Darwin, in the Northern Territory, there are domes with a scatter
of blocks all but covering them, and near Gokomere, in central Zimbabwe, domes
with marginal developments of blocks (Fig. 5.34). In central Namaqualand and
Namibia there are hills that are half dome and half nubbin (Fig. 5.35), and so
on - many examples of such intimate coexistence of forms can be seen in the
field with elements of different types of inselberg developed on the same rock
mass.

To some extent they can be explained in terms of structure. Thus in the
Namaqualand examples just cited, the domed elements are coincident with granite,
the blocky ones with gneiss. On the other hand, there are many morphological
contrasts that have no such obvious structural basis.

(i) Nubbins

Nubbins are block- or boulder-strewn residuals. They are roughly conical,
though many are flat-topped. Nubbins are particularly common in warm, seasonally
humid climates, in regions like the tropical monsoon zones of northern Australia
(northwest Queensland (Fig. 1.17); the Darwin area; the Pilbara); in Hong
Kong and the adjacent Territories (e.g. the Shatin area). They are also present
in arid lands, as, for instance, at Alice Springs and in the Llano Dome of
central Texas, in both areas in wet sites in intermontane basins. They are
developed in valley floors in the Swakoprivier valley, south of Karibib in
central Namibia (Fig. 5.19), and the Witrivier valley of central Namaqualand.
In some areas, as, for example, the Mohave Desert of southern California

Castellated structures surmounting large radius domes, Fig. 5.32 (above) at
the Devil's Marbles, N.T., and Fig. 5.33 (below) hollowed by tafoni development,
on Remarkable Rocks, Kangaroo Island, S.A.

(Oberlander, 1972), they are regarded as inherited from former humid conditions.
In Angola, too, where nubbins are well developed (Amaral, 1973), there are
grounds for suggesting that conditions were formerly wetter than they now are
(Wayland, 1953).

In areas where nubbins and bornhardts are found side-by-side, as, for
instance, in the western Pilbara, the two types are found to be constructed of
the same sorts of material so that here the contrast between dome and nubbin is
not due to petrological characteristics. Sheet structure is well developed but
so are orthogonal joint sets. For instance, in the Mt Bundey area, near Darwin,
the fresh granite is seen in quarry exposures to be subdivided into cubic and
quadrangular blocks (Fig. 2.19), yet arcuate joints are dominant in the domed

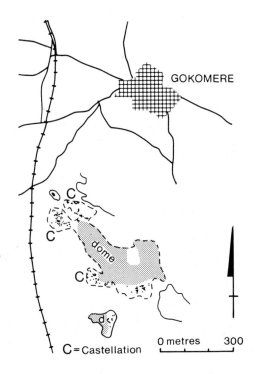

Fig. 5.34. Plan of Gokomere inselberg, Zimbabwe, showing blocky margins.

Fig. 5.35. Inselberg built partly of granite, partly of gneiss, and forming a half dome and half nubbin, near Witrivier, Kamiesberge of central Namaqualand, R.S.A.

Fig. 5.36. Sheeting fracture and blocks of disintegrated sheet structure, Mt Bundey, N.T.

features that underlie the outer jumble of blocks and boulders (Figs. 5.36, 5.37, 5.38).

Compared with the structures on which castle koppies are developed, however, the radius of these domes is small, and nubbins appear to be domes, the outer shells of which have been broken down to blocks and boulders through the exploitation of the fan and cross joints referred to earlier. Continued disintegration of the shells eventually reduces the mass to low, large radius domes or platforms (Figs. 5.39 and 5.40).

There is considerable evidence that nubbins are developed by subsurface weathering under warm, humid conditions. The flat crests of nubbins in northwest Queensland and at Alice Springs are readily correlated with early Tertiary palaeosurfaces (Twidale, 1956; Mabbutt, 1965) of which they are thought to be part, possibly in etch form, and below which deep differential compartment weathering took place under warm, humid or subhumid conditions. Handley (1952) has shown that what he called the tors (which are in reality nubbins) of Tanganyika evolved beneath the African land surface of early Cainozoic age. Those of the western Pilbara developed beneath the (?Cretaceous) Hamersley Surface, and so on - many nubbins are associated with, and occur below the level of, palaeosurface remnants.

Fig. 5.37. Above, nubbin with sheet structure visible, western Pilbara, W.A., and Fig. 5.38, below, nubbin with exposed domical core at northern end of Paulshoek, in the Kamiesberge of central Namaqualand, R.S.A.

The subsurface weathering of the outer shells of the domed masses has resulted in the alteration of rock along fractures and the development of core-stones with pitted surfaces set in grus within the limits of the sheet structure. Examples of such forms still in process of exposure can be observed in Angola (Amaral, 1973) at several sites on Eyre Peninsula, and at Paarlberg, in Cape Province, South Africa (Figs. 4.27 and 4.28).

For these reasons it is suggested that most nubbins have their origin in the subsurface and evolve particularly well in warm, humid climates, where weathering is sufficiently aggressive to cause the blocky disintegration of the outer shell of shells of the convex-upward masses of still fresh rock. Some blocks and boulders doubtless result from the disintegration of sheet structure after exposure, but most of it probably occurs below ground. After the lowering of the plains and the exposure of the residuals, the continued breakdown of the blocks and boulders and particularly the evacuation of the interstitial grus causes them to become disarranged as they tumble downslope under gravity. In

Fig. 5.39. Low dome, Mt Bundey area, N.T.

Fig. 5.40. Low dome near Nanutarra, western Pilbara, W.A., with, in foreground fretted, weathered blocks impregnated with oxides of Mn and Fe, and in background, uplands in sedimentary rocks with summit surface (part of Hamersley Surface of ? Cretaceous age).

this way, parts of the inner dome are revealed. Within fields of nubbins located in humid areas, however, there are few bare domes, suggesting that, though the blocky veneer disintegrates, the core mass also continues to decay, so that, although the size of the nubbins is decreased, the block-strewn morphology is retained (Fig. 5.41).

Thus, nubbins are domes based on conformable structures of small radius and

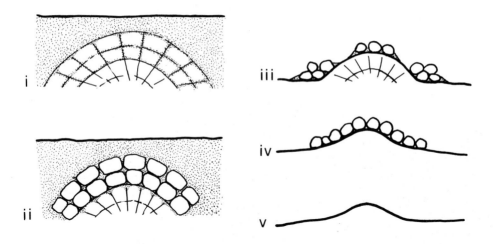

Fig. 5.41. Schematic development of nubbins.

the outer shells of which have been broken down into quadrangular blocks by
subsurface weathering under warm, humid conditions. Coastal nubbins are exceptional in that the sheet structure has been broken down under weathering and
wave attack.

(ii) Castle koppies

Castle koppies have much in common with bornhardts. Both types of inselberg
are developed in massive bedrock. They are not devoid of fractures, but the
partings are relatively few in number and are tight. Both are subdivided by
both orthogonal and sheeting fractures but, in the koppies, the former are
dominant, whereas the latter are more prominent in the domical forms. Both
occur in multicyclic landscapes (Fig. 5.42). Both display evidence, in the form
of zones of flares well above the present plain level, of phased development
(Fig..5.43).

According to Demek (1964a and b) the koppies of Bohemia are *Fernlinge* remaining after scarp retreat, and Watts (1979) interprets the castellated residuals
of the Canadian North in the same light. Ritchot (1975, pp. 253 *et seq.*) views
castle koppies as part of a sequential series, the end member of which is the
bornhardt. He argues that the marginal areas of structural domes persist as
landscape features longer than do the central zones, because the latter are in
tension.

Godard (1977, p. 101) considers that castellated inselbergs are either due
to frost shattering or are large residual masses remaining after differential
subsurface weathering (giant core blocks) or are the cores of inselbergs remaining after either scarp retreat or differential rock disintegration; but the
discussion is carried no further.

Fig. 5.42. Castle koppie at Devil's Marbles, N.T., with, in background, remnant of Early Cretaceous palaeosurface preserved on sandstone ridge.

Domed and castellated inselbergs developed in similar rock types coexist in the Salisbury Mrewa-Marandellas area of Zimbabwe and elsewhere. Again, at many sites castellated forms rest on large radius domes, but in most instances the bedrock appears to be similar, if not identical.

The angular morphology of castle koppies undoubtedly reflects either massive orthogonal fractures or a well developed vertical or near-vertical foliation. Thus on Dartmoor, the Massif Central, and in many other places, castellated granite inselbergs are associated with orthogonal jointing (Figs. 5.44, 5.45 and 5.46). But not all granites with well developed orthogonal joint sets give rise to castle koppies; the Mt Bundey area provides one example that has already been mentioned and, in any case, if the association were invariable, castle koppies would be very widespread, whereas they are, perhaps, the least common of the three inselberg forms discussed here.

Many of the koppies of southern Africa (Zimbabwe, Malawi, Botswana) are developed on gneisses with a well developed, steeply-dipping, widely-spaced foliation. There are many examples of such gneissic koppies in the eastern Mount Lofty Ranges and in the southwest of Western Australia, e.g. in Castle Rock. On the other hand, many bornhardts are developed on gneissic rocks and many castellated hills are formed in granite. In the Kamiesberge of central Namaqualand, for instance, foliation has been exploited by subsurface weathering, so that the ribbed lower slopes stand in contrast with the rounded hill crests

Fig. 5.43. The bounding slopes of Castle Rock, in the Mt Manypeaks complex near Albany, southwest of W.A., are flared, suggesting phased exposure.

(Fig. 5.47). Well-foliated gneiss is not, in itself, a guarantee of castellated form. The Reynolds Range of central Australia, for example, consists of granite gneisses on which are developed domes with a discontinuous veneer of platy blocks. Foliation is clearly well developed and is exploited by weathering to give penitent rocks (Fig. 4.11 (b)). Yet there is here no suggestion of castellated morphology (Fig. 5.25).

Thus, though structure is a significant factor in the genesis of castle koppies, it is not an overriding one: evidently the structural base has been exploited in particular ways and conditions for the angular form to evolve.

Godard (1977, p.99) refers to castle koppies as *inselbergs de poche*, indicating that the castellated forms are small compared with bornhardts. They are neither as high nor as areally extensive as their domed counterparts in any given area. Thus in the Wudinna area of northwestern Eyre Peninsula the smallest of the residuals is the Cottage Loaf which is a lone castellated form. It stands on a plinth or low platform that is, in fact, part of a large radius dome, and in this respect is similar to the Devil's Marbles (central Australia), Remarkable Rocks (Kangaroo Island, South Australia), several remnants on the Mount Manypeaks complex (near Albany, Western Australia), and some in the Tassili Mountains of the central Sahara (Rognon, 1967). That many koppies are located on the crests of large radius domes suggests that they are the last remnants of massive sheet structures, the marginal areas of which have been worn away. Only the crestal zones are preserved, suggesting that there has been strong marginal attack, though, because of the gentle inclination of the fracture-controlled slopes, there is not the same tendency for blocks to tumble downslope and

Fig. 5.44. Castle koppie in the Tosa Gargantillar (1700 m) in the high Pyrenees.

Fig. 5.45. Castle koppie at Haytor, eastern Dartmoor.

become disarranged as there is with the small-radius nubbins.

Subsurface weathering in the piedmont zone finds its most dramatic expression in flared slopes which are developed not only on granite but also in other

Fig. 5.46. The Ranc de Bombe (1550 m) is a castle koppie in the Margeride southwestern Massif Central.

Fig. 5.47. Rooiberg, in the Kamiesberge of central Namaqualand, showing ribs related to foliation planes, more prominent beneath old land surface (shoulder S) where structural weaknesses were subjected to moisture attack.

lithological environments. They are a particular form of the weathering front developed in the scarp foot zone (Twidale, 1962, 1967, 1971a, 1976b - see also Chapter 9).

Repeated alternations of subsurface weathering and lowering of the plains have allowed the development of several flares in some slopes (Fig.5.43). Some such multiple flared slopes, like that on the eastern side of Ucontitchie Hill, stand up to thirty metres above the plains and expose the truncated ends of several sheet structures (Fig. 2.28). Clearly, there has been a reduction in the area of the hill, and the width of the platforms exposed between the base of the flare and the steep plunge of the bedrock that mark the fracture-controlled edge of the massive compartment indicates the amount of marginal trimming that has taken place as a result of scarp foot subsurface weathering. At some sites it amounts to some tens of metres, which is insignificant in terms of the origin of the inselberg, but is sufficient to transform the convex-upward flanks of a dome into a steep-sided cliff, and, if continued, a dome into a koppie.

Similar processes are and have been at work in scarp foot zones without producing flared footslopes (as, for example, in well-jointed or fissile rock), but they are responsible for the basal steepening of scarps and for the peripheral disintegration of sheet structures, resulting in the formation of marginal fields of blocks and boulders such as have developed at Waulkinna Hill, on northern Eyre Peninsula, and at Gokomere (Fig. 5.34) and Domboshawa, both in Zimbabwe.

Several conditions are especially conducive to pronounced marginal attack. In arid and semiarid lands there is a tendency to steep slopes due to the contrast between active weathering in the moist subsurface and the stability of the dry, and hence stable, exposed surfaces. It may well be that, whereas nubbins are initiated wholly in the subsurface, koppies tend to evolve where the crests of dome structures are exposed (Fig. 5.48). Local wet sites which, for structural reasons, are of long standing produce pronounced marginal weathering as, for instance, at the Devil's Marbles which are located in an enclosed basin flanked by sandstone ridges of an anticlinal structure (Figs. 5.42 and 5.49). Long periods of landscape stability such as are associated with the various land surfaces in such interior sites as Zimbabwe, a country rich in koppies, allow even modestly aggressive weathering processes to have marked effects. All that is necessary is that the upper part of the residual be either exposed or located in the drier, near-surface zones of the regolith, while lower parts are affected by moisture in the deeper regolith. Thus the Devil's Marbles were initiated by differential weathering beneath a land surface of probable Early Cretaceous age (Twidale, 1980c). Last, vertical or near-vertical cleavage or other fractures not only allows moisture to penetrate into the rock but also imposes a measure of structural control on the form of the weathering front.

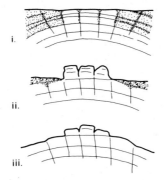

Fig. 5.48. Schematic development of castle koppies.

Some castellated forms are, thus, seemingly explicable in terms of peripheral
weathering of domical masses in the deeper regolith, resulting in the isolation
of massive, centrally-located blocks delineated by orthogonally arranged fractures. Eventually, only isolated blocks remain and, in due course, they, too,
are disintegrated, leaving a low dome or platform of limited areal extent. Thus
the penultimate stage of the destruction of bornhardts, regardless of whether
the intervening forms are nubbins or koppies, involves a large radius dome or
platform; Ritchot (1975) is correct to cite bornhardts as a late-stage product
of an inselberg cycle, but, more importantly, they are also the starting point
for nubbins and castellated forms.

Castellated inselbergs are well developed in areas like Lapland and Newfound-
land (Schrepfer, 1933); The Rocky Mountains of Colorado and Wyoming (Eggler *et*
al., 1969; Cunningham, 1969) and Mongolia (Dzulinski and Kotarba, 1979) that
are seasonally very cold and have in the recent past been colder and even
glaciated. They are also commonplace in upland settings that are seasonally
cold and experienced either glaciation or pronounced frost action during the
Pleistocene: areas like Dartmoor (Linton, 1955), the Rocky Mountain high plains
in Colorado and Wyoming (Eggler *et al*., 1969), the Bohemian Massif (Demek, 1964a
and b), the Pyrenees, the Massif Central (Fig. 5.46), the uplands of Tasmania
(e.g. Caine, 1967), the Snowy Mountains and adjacent uplands in southeastern
Australia and the New England Tableland of N.S.W. (Leigh, 1967, 1970).

Like bornhardts and their castellated congeners in warmer lands, these cold
landforms are associated with palaeosurfaces beneath which there may have been
intense and prolonged weathering. Thus the koppies of the Rockies stand upon
the Sherman Surface (Eggler *et al*., 1969), and the residuals of the high
Pyrenees are associated with a high plain.

Linton (1955) attributed the Dartmoor tors to differential compartment
weathering under humid, tropical conditions and subsequent exposure of the
upstanding residuals by solifluxion and cold-climate processes generally.

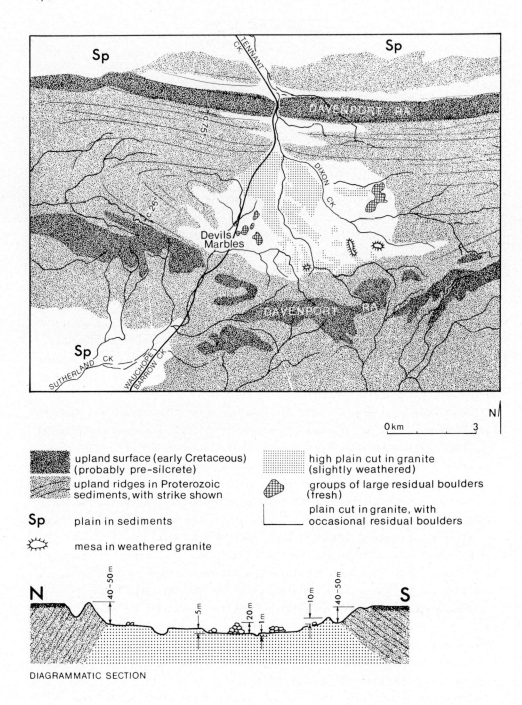

Fig. 5.49. Morphological map and section of Devil's Marbles and environs, N.T.

Demek (1964a and b) reached comparable conclusions regarding the koppies of the Bohemian Massif. Leigh (1970) considered that the koppies of New England (north-eastern New South Wales) were initiated in the subsurface: and so on. In these terms, the only essential differences between nubbins and castle koppies are, first, that the former are developed on small-radius domes and the latter on large-radius structures; and, second, that the nubbins evolved in the deeper subsurface, whereas the crests of koppies were either exposed or located in the shallow subsurface.

On the other hand, there seems no reason why frost action should not have continued, perhaps working in the earlier developed regolith, to cause further steepening of the masses of fresh rock protruding into the regolith, much as suggested by Palmer and Nielson (1962). Alternatively, freeze-thaw action in the scarp foot zone during the gradual exposure of the residuals could produce the cliffed bounding slopes. Either of these mechanisms accounts for the steeper bounding slopes of the castellated forms.

The residuals that survive on high plain remnants in such areas as the Pyrenees do so because the host masses on which they have evolved are in stress and thus resistant. The weaker zones were weathered and subsequently have been eroded, in some instances deeply, by rivers and glaciers or both.

Castle koppies are of varied origins, as Godard (1977) suggested. They form in both the tropics and in cold lands, and are domes that have been subject to marked marginal weathering, most probably in the comparatively shallow sub-surface; it is likely that the crests of the structures were exposed during the subsurface marginal attack, so that the upper surface remained stable or was weathered only very slowly.

Bornhardts, nubbins and castle koppies are genetically related forms, the two last named being derived from the marginal subsurface weathering of domed forms. Nubbins and castle koppies are the shattered remnants of bornhardts.

The basic form, the bornhardt, is a structural feature developed on masses that are compact by virtue of their being in stress. They are characterised by and owe their domical shape to the development of sheet structure as a result of lateral compression. Nubbins and castle koppies, on the other hand, though strongly influenced by structure and also found in multicycle landscapes, are in some measure morphogenetic features. Nubbins are best developed in humid, tropical areas as a result of the superficial disintegration of the outer shells (sheet structures). Castle koppies, on the other hand, are due to lateral or marginal weathering, also in the subsurface, and under various climatic condi-tions (tropical, arid or semiarid, arctic or subarctic). The two derived forms probably reflect both the geometry of the domed structures - small radius, in the case of nubbins, large, in the case of koppies - as well as the degree of exposure of the residuals, for, whereas the nubbins have been weathered over their entire surfaces, the castellated forms appear to have been attacked from the sides, suggesting that the upper part was exposed, or at least close to the surface, during marginal attack.

But weathering eventually reduces both nubbins and koppies, though the latter especially are very durable. What remain are small domes which are frequently scarcely more than low, convex-upward platforms. They are nevertheless geneti- cally related to bornhardts, so that a cycle of inselberg development is discernible.

The dome structure is the starting point of an evolutionary sequence that can follow varied paths to attain the same final destination. All three dominant inselberg forms are initiated in the subsurface. Bornhardts begin life as domical masses protruding into the regolith, and their domed form is maintained through exposure and subsequent evolution. They are compact forms with the least surface area in relation to their volume. But in some conditions the dome struc- tures are attacked differentially in the subsurface, producing block-strewn and castellated masses which, when exposed, become nubbins or castle koppies. The latter are weathered to expose the low domes which, if and when the adjacent plains are lowered, will again become inselbergs - domed, block-strewn or castellated, according to their structure and environment.

CHAPTER 6

ALL-SLOPES TOPOGRAPHY

Some granite uplands are neither domed nor block-strewn nor castellated, but rather consist of sharp-crested ridges bounded by essentially rectilinear slopes. The latter vary in inclination from area to area but they are characteristically steep. There are no significant areas of flat land either on ridge crests or in valley floors, and on this account such topography is called *all-slopes*.

A. DISTRIBUTION

Regions of all-slopes developed on granite have been described from the Sinai Peninsula (Hume, 1925), the Peruvian Andes (Kinzl and Schneider, 1950; Gerth, 1955 - see Fig. 6.1), the Pyrenees (Barrère, 1952), the northern Flinders Ranges, South Australia (Twidale, 1964 - see Fig. 1.20), the uplands of Papua New Guinea (Pain and Ollier, 1981), southern Greenland (Soen, 1965 - see Fig. 6.2), the Sierra Nevada of California (Fig. 1.1), and the Karkonosze Mountains of southern Poland (Dumanowski, 1968). Extensive areas of all-slopes are developed in the Serra do Mar, Serra dos Araras and the Rio de Janeiro area (Tijuca Massif) of southeastern Brazil (Fig. 6.3) around the adjacent, but higher, Agulhas Negras (Fig. 6.4). There are also spectacular examples of especially steep-sided, almost acicular, all-slopes in the eastern Sierra Nevada, for instance, around Mt Whitney, in the Cathedral Rocks of the Yosemite Valley, and in The Needles of South Dakota.

All-slopes topography is not confined to granitic rocks, similar assemblages having developed, for example, on shales in the southern Flinders Ranges west and northwest of Wilmington, in the Pertnjara Hills in central Australia, and in

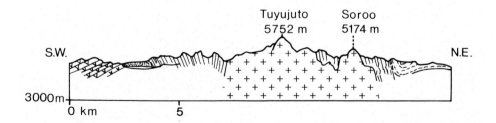

Fig. 6.1. Section through part of the Peruvian Andes (after Gerth, 1955).

Fig. 6.2. All-slopes topography in granite, Sermasoq, southern Greenland (Oen Ing Soen).

Fig. 6.3. All-slopes topography in the Serra do Sol, near Rio de Janeiro, southeastern Brazil.

Fig. 6.4. All-slopes topography in syenite, Agulhas Negras, at an altitude of 2787 m in the Serra da Montiqueira of southeastern Brazil.

folded sediments and metasediments in the Snowy Mountains of southern N.S.W. and northeastern Victoria.

B. ORIGINS

All-slopes topography has evolved in various lithological and climatic set-tings - periglacial or nival, humid tropical, and hot deserts. Several possible explanations come to mind.

The development of all-slopes topography in cold uplands suggests that the climatic conditions experienced now and in the recent past are conducive to the formation of the rectilinear forms that constitute this particular type of topography. At present it is frost shattering that is shaping the slopes, but during colder phases of the Pleistocene, glaciers were active. In these condi-tions cirque and valley glaciers with frost-riving active on the upper exposed slopes could have produced all-slopes topography with the combined action of ice and frost causing the recession of bluffs, the intersection of opposed scarps and the formation of *arêtes* and pyramidal peaks or horns (Fig. 6.2). After the waning of the glaciers, frost shattering continued, further wearing back the steep, frost-riven cliffs, and causing them to be breached in some places and producing cores of screen that blankets lower slopes elsewhere (Fig. 3.11). Where the opposed scarps remain some distance apart, remnants of palaeoplains survive (Fig. 6.5), but where they are closely juxtaposed, vertical fractures have been exploited, leaving tall, isolated pillars known in the Pyrenees as

Fig. 6.5. Remnant of an (?) early Cainozoic planation surface preserved on the Crestes de Gargantillar, in the Andorran Pyrenees.

Fig. 6.6. *Gendarmes*, turrets of granite in the Crestes de Gargantillar, Andorran Pyrenees, and presumably so called because they stand tall and upright.

gendarmes (Fig. 6.6).

In general terms, the exposed upper slopes are worn back by nival action and recede as envisaged long ago by Fisher (1872) and others. The lower slopes are protected by angular scree and, in the long term, the slopes are lowered, but as long as the cliffs persist, they maintain a constant inclination during recession induced by frost action (Twidale, 1959). Clearly, however, nival processes do not provide a complete explanation of all-slopes topography which is found in areas that have suffered neither glaciation nor significant frost action during the relevant period of landscape development: thus the Sinai, the northern Flinders Ranges, and the coastal uplands of southern Brazil (*pace* Agassiz, 1865) have not been subject either to late Cainozoic glaciation or strong frost action. Again, not all periglacial or nival areas underlain by granite have been shaped into angular forms. As is the case in some arid regions, domes are interspersed with all-slopes. Thus in southern Greenland, as in the northern Flinders Ranges, the two types of terrain occur side-by-side, reflecting the juxtaposition of massive and well-jointed rocks. On the other hand, the massifs of southwestern England (Dartmoor, Bodmin Moor, etc.) and the Massif Central were nival if not periglacial during the Pleistocene, yet neither displays all-slopes topography though some of the valleys cut in the margins of the uplands display rectilinear side slopes; all that is needed to produce all-slopes is a greater stream density.

All-slopes topography can be interpreted as reflecting the attainment of equilibrium conditions on slopes eroded in effectively isotropic or homogeneous materials. Ahnert (1967) has provided an idealised scheme in which the results of variations of waste supply (weathering) on slopes, removal of that waste from the slope and from the slope toe or base are considered. All-slopes can be seen as a response to structural homogeneity and particularly the weathering of granitic rocks that are so fissured that disintegration and erosion are essentially uniform over the slope. If it is allowed that baselevel permits deep incision, and that the climate is either arid or semiarid (or at least seasonally arid), then it can be suggested that as streams incise their beds weathering produces just enough debris to be transported down the slopes and from their bases, so that linear graded slopes develop and are maintained. Although weathering and erosion take place, slope morphology remains constant.

A variation on this theme was suggested in general terms by Twidale (1968, pp. 323-325) who pointed out that in tectonically active regions, equilibrium between uplift, stream incision and slope development could result in true independent forms. Such an explanation is advocated in respect of the Papua New Guinea granitic all-slopes terrain (Pain and Ollier, 1981) and may be applicable in other unstable regions.

Thus in the northern Flinders Ranges, which have been uplifted in late

182

Cainozoic times, the main, older mass of granitic rocks is evenly dissected. The rocks are essentially fresh, though lower slopes are weathered. On the slopes, which average 35°-40° inclination, bedrock is everywhere exposed. Clearly, waste removal at least equals the production of granite sand by weathering and, equally, the streams, though flowing only episodically, are nevertheless capable of transporting whatever debris is carried under gravity and by wash to the foot slope, for there are no significant talus accumulations. The rectilinear slopes appear to be in equilibrium and to be eroded uniformly. They may be receding, though alternatively these bare rock slopes may be essentially stable in the prevailing aridity. But in the same region there are steep-sided domes or sugarloaves, such as the Armchair (Fig. 1.20), which are clearly an expression of structure and which are bounded by comparatively low-angle cones, so that the slopes in profile are faceted, with bluff and debris slope clearly developed.

Similarly, in the Rio de Janeiro area there are many ridges and ranges that consist of knife-crested uplands, but again there are many examples of *morros* or sugarloaves rising from such forms (Fig. 6.7). The uplands around Rio include what can be interpreted as various stages in a postulated sequence involving scarp recession. The domes surmounting conical hills represent a penultimate stage in the formation of all-slopes. Continuation of the process would lead to the elimination of the residuals and of the faceted slopes, and the production of all-slopes terrain similar to that displayed in the nearby Serra do Sol

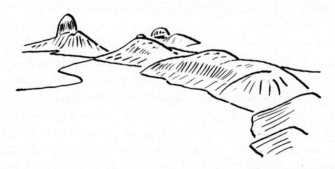

Fig. 6.7. Sketch of part of the Rio de Janeiro area, showing faceted slopes, the bluffs being in fact the flanks of the steep-sided morros (drawn from a photograph in Lamego, 1938).

(Fig. 6.3). Here, the forms are underlain by deeply weathered rock, but whether the slopes are excavated in previously weathered granite or whether the deep weathering has developed in consequence of the reduced rate of slope sculpture is not known. The steeply inclined (*circa* 20°-25°) slopes carry an almost

uniform veneer of red soil. Clearly, weathering and soil formation have outpaced erosion, but it appears that, at every point on the hillslope, erosion
keeps pace with weathering, and that the drainage system has evacuated debris as
it has arrived at the hill base. In this and other, similar, areas, equilibrium
between weathering, erosion and evacuation appears to have been attained. The
system has, however, been disturbed by road construction and by clearance of
vegetation, with the result that landslips and slides have developed.

Isolated examples of the all-slopes, or near all-slopes, forms can be interpreted as late-stage cyclic developments in the context of scarp retreat. It
can be argued that in arid conditions, especially, basal weathering and erosion
exceed that on upper slopes, and that in these circumstances slope retreat prevails (Tricart, 1957; Twidale, 1960). That steep slopes are maintained and
that slopes recede can be attributed to the comparative desiccation of upper
slopes and the concentration of moisture on lower slopes, and particularly at
the scarp foot. That granite is susceptible to moisture attack and relatively
stable when dry has been noted by several writers, and the effects of moisture
concentration at the base of slopes by many writers (e.g. Peel, 1941; Twidale,
1967). Thus it can be suggested that because it is dry the granite of upper
slopes and crests acts as a stable and resistant material in contrast with the
weathered and weak granite of lower slopes. In such situations, erosion of the
lower slopes tends to outpace that on upper areas and the slopes tend to be
maintained at the steepest inclination commensurate with stability (Tricart,
1957; Twidale, 1960). Where streams are relatively widely spaced, faceted
slopes come to border remnants such as mesas and buttes (Figs. 6.8 and 6.9),

Fig. 6.8. Sketch of *mamelon* or butte in pegmatite, Fort Lamy region, western
Sahara (after Foureau, 1905).

as in the region of the Devil's Marbles, Northern Territory, in parts of northwest Queensland, and in the western Sahara (Foureau, 1905, p.610). Where,
however, streams are closely spaced, the bluffs are either eliminated or never
existed, and the rectilinear debris slopes intersect, resulting in the development of true all-slopes topography.

In most temperate regions (such as southwestern England, central France) the upper slopes are never dry enough to act as caprocks and, for this reason, plus the permeability of the weathered rock, and the consequent wide stream spacing, all-slopes did not evolve.

Both the nival and the explanations resulting in slope equilibrium and scarp retreat involve epigene processes but in some cases, and particularly in the northern Flinders Ranges, it is possible that the all-slopes relief reflects patterns of earlier subsurface weathering. The granitic all-slopes topography of the northern Flinders Ranges occurs beneath prominent remnants of a palaeoplain (Fig. 1.20) of Early Cretaceous age (Woodard, 1955; Twidale, 1980d). There may have been fracture-controlled differential weathering beneath the palaeoplain with deeper alteration beneath the valley floors than beneath the divides (Fig. 6.11). Thus, with reference to the Flinders region, when streams were rejuvenated following renewed uplift of the Ranges during the Miocene (Callen and Tedford, 1976), the regolith could have been evacuated to expose the weathering front as an irregular, all-slopes topography, which was then smoothed by the operation of various slope processes.

Thus all-slopes topography may evolve in different ways in different environments. Everywhere it reflects equilibrium conditions on slopes, and it is essential to have structural or weathering conditions that are conducive to the

Fig. 6.9. Faceted slope in granite, Isa Highlands, northwest Queensland (Div. Land Res., C.S.I.R.O.).

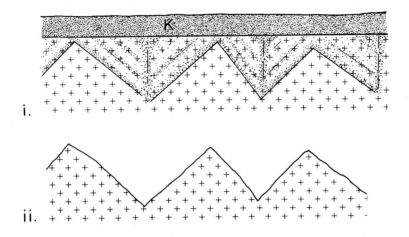

i.

ii.

Fig. 6.10. Suggested development of all-slopes as an etch surface, northern Flinders Ranges, South Australia (K - Cretaceous).

maintenance of an upper slope. In some areas it is the dryness of the rock exposed in upper slopes, elsewhere strong basal slope attack, and yet elsewhere a combination of these two conditions. Where the bedrock is intensely weathered in such a way that it is friable and lacking strength, steep slopes cannot be maintained, and plains are formed bordering any residual remnants. Thus on Dartmoor and northwestern Eyre Peninsula there is an abrupt transition between hill and plain, and no debris slopes, and basal attack is too efficient in some arid areas so that the available debris has been evacuated. Elsewhere, however, the debris protects the underlying bedrock, and towers, turrets or domes are fringed by relatively low-angle cones. In humid warm areas like the Rio region the vegetational veneer no doubt assists in the preservation of these debris slopes.

CHAPTER 7

GRANITE PLAINS

Despite the understandable interest evinced in the positive relief features
developed on granite and discussed in the previous several chapters, plains are,
measured by their areal extent, by far the most characteristic landform developed
on granitic bedrock. Plains are indeed an essential component of such well-known,
and typically granitic features as *Inselberglandschaften*, for it is the contrast
between the virtually featureless sweeping plains and the steep-sided residuals
that endows the landscapes with their dramatic aspect. Yet the landform assemblage is named after the upstanding, but areally minor, features.

That plains are widely developed in granite is not surprising. Granites in
general are particularly vulnerable to water attack. They commonly contain mica,
which is readily susceptible to alteration and, as with a chain, a rock is only
as strong as its weakest link. Once the defences are breached by the weathering
of one constituent of reasonable abundance, water can all the more readily enter
the rock mass and continue its attack on relatively stable minerals. Moreover,
and as has been mentioned previously, granodiorite is not only by far the most
common of the granitic rocks, but consisting, as it does, of over 40% feldspar,
predominantly plagioclase, which is readily susceptible to moisture attack and
alteration to clays, it is, according to Lagasquie (1978) the weakest, the rock
most vulnerable to weathering, of all those in the granitic domain.

Moreover, granites form the foundations of the continents. Many are of great
antiquity, and have, moreover, been either exposed or in the groundwater zone
for a long time. Again, it is increasingly apparent that for very long periods
of geological time warm humid conditions prevailed over large areas of the continents. Many aspects of the modern deserts, for example, are relic from such
past periods of humid, warm conditions, in which granitic rocks are prone to
rapid, intense and deep weathering. The high plains of Dartmoor (and other
granite massifs of the southwest of England), of the Hercynian massifs of
continental Europe, and of the Monaro district of southeastern N.S.W., are cut
mainly in granite. These plains of low relief, though not now located in the
middle latitudes, probably evolved in part at least under warm and humid conditions that prevailed during much of the Mesozoic and all of the earlier Cainozoic
(see e.g. Büdel, 1959, 1963; also Linton, 1955; Kieslinger, 1960; Demek,
1964a and b)

It is no accident, either, that many deep weathering profiles developed on granitic rocks are reported from the various shield lands (see Chapter 2) which, though subject to epeirogenesis, have nevertheless been relatively stable for long periods, or that plains are well and widely developed on older granite masses. Plains are characteristic of the various Precambrian shield areas and of various Palaeozoic massifs, yet they are also well developed on Cainozoic bodies that have been exposed only for relatively short periods: palaeoplain remnants are well developed, for instance on granitic rocks in the Yosemite region of the Sierra Nevada of California, and on older granitic masses involved in Cainozoic (Alpine) orogenesis, as, for instance, in the Pyrenees (Fig. 6.5).

A. BURIED AND EXHUMED PLAINS

Many granitic plains have been recognised in unconformity in the stratigraphic column and some have been *exhumed* to form part of the contemporary landscape. Thus, as an example of the first of these still-buried types, in the area south of Cape Town the Table Mountain Sandstone rests on a planate surface cut in Pre-cambrian granite. The unconformity is well exposed in coastal sections and road cuttings (Fig. 7.1), and is indeed followed by the main road linking Cape Town with the Cape of Good Hope over considerable sectors. The unconformity is of quite remarkable evenness. In similar fashion, the Torridonian sandstone in northwest Scotland rests on what appears to be an old inselberg landscape, developed in Lewisian gneiss, and consisting of domical residuals surrounded by pediments (Williams, 1969) that were blanketed by the younger sediments but which have been partly re-exposed and glaciated. Cowie (1961) had earlier des-cribed what he termed a Precambrian peneplain developed in crystalline rocks and partially exhumed from beneth Proterozoic sediments (Fig. 7.2), and Ambrose (1964) has identified similar surfaces of low relief preserved in crystalline terrain, but resurrected from beneath Proterozoic and earlier Palaeozoic cover in several parts of the Canadian Arctic.

B. ETCH PLAINS

Where the weathered mantle has been stripped away by rivers, the weathering front has been exposed as an *etch plain* (see e.g. Jutson, 1914; Wayland, 1934 - also Fig. 7.3). Such stripping is facilitated by lowering of baselevel but is not necessarily an indication of such baselevel movement for rivers could develop lower gradients in debris reduced in size by weathering and could thus cut into the regolith.

Much of the Labrador Peninsula is a high plain eroded largely in granitic materials (see Tanner, 1944). It has been modified by Late Cainozoic ice sheets and, in particular, has been stripped of most of its regolith, so that it is of etch character. In addition it may also in part be exhumed.

The extensive plains cut in granite in the southwest of Western Australia are

188

Fig. 7.1. Unconformity between granite below and Table Mountain Sandstone above, south of Cape Town, R.S.A.

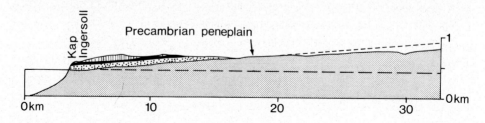

Fig. 7.2. Exhumed peneplain eroded in granitic rocks, northern Greenland (after Cowie, 1961).

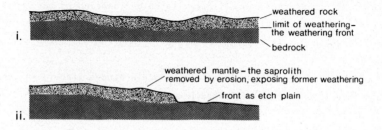

Fig. 7.3. Stages in the evolution of an etch plain.

also of etch character. The lower surface has been called the New plateau (Jutson, 1914, 1934), though it is more appropriately called a high plain. Nonetheless, there are remnants of an older lateritised plain preserved as plateaux, mesas and buttes and standing above the level of the New surface (Fig. 7.4). A lateritic weathering profile is preserved in these residuals and well-exposed in the bounding scarps that delimit them. The base of the regolith is roughly coincident with the level of the New plateau (Fig. 7.4) so that the latter is of etch character (Jutson, 1914, 1934; Mabbutt, 1961b). An etch surface that is in part developed on granitic and gneissic rocks has been identified in the eastern Mount Lofty Ranges of South Australia (Twidale and Bourne, 1975c), and many relatively minor platforms (see below) are of similar character.

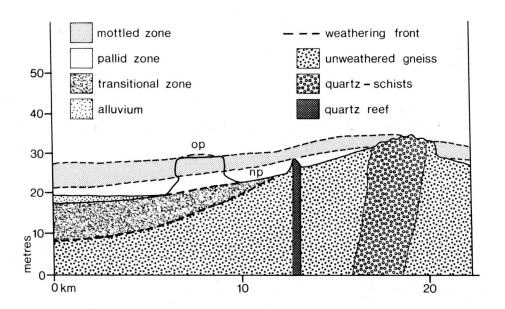

Fig. 7.4. Relationship between the older lateritised surface or plateau (o.p.), weathering zones and the New plateau (n.p.) in the Wiluna area, central W.A. (after Mabbutt, 1961b).

C. PLAINS OF SUBAERIAL (EPIGENE) ORIGIN

Plains can be regarded as the granitic norm. They are well represented in tropical and subtropical lands, though they are by no means restricted to them. Though bedrock plains cut in fresh granite by marine agencies are likely to be of only limited extent (see Chapter 1: Fig. 1.7), there is no reason why wave action should not extensively erode weathered granite, especially if there were a secular relative rise of sealevel. Some plains preserved in unconformity may be of this type, or at any rate be epigene forms retouched by marine action during oceanic transgressions.

Plains developed by epigene processes on granitic rocks vary in both extent and morphology. The gently inclined pediments are of limited areal extent, whereas rolling or undulating peneplains occupy huge areas; in addition, certain other plains are unbelievably flat and featureless.

(i) Pediments

Pediments are gently inclined, cut bedrock surfaces located in the scarp foot zone (Fig. 1.25). Though many pediments carry a regolithic veneer, the surface form reflects the slope of the bedrock surface. This is the essential difference between pediments and alluvial fans, for the form of the latter is a consequence of deposition, and the inclination of the surface is a function of the gradient of the streams responsible for the transportation and deposition of the debris. On granitic rocks the slope of pediments varies between $\frac{1}{2}^{o}$ and 7° but they are typically inclined at $\frac{1}{2}^{o}$-$2\frac{1}{2}^{o}$ with respect to the horizontal. Most are gently concave upward (Fig. 7.5), many are rectilinear, and a few are convex. For

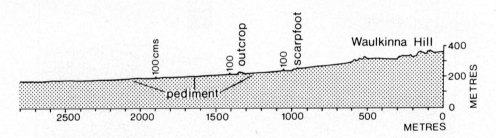

Fig. 7.5. Section through Waulkinna Hill and associated pediment, northern Eyre Peninsula. Depths of sand cover shown in cm.

example, in the Georgetown area of northwest Queensland streams are actively incising and the toes of the pediments are clearly convex upward. But whatever their geometry, pediments meet the backing escarpment in an abrupt break of slope called the piedmont angle or pediment nick (Rahn, 1966; Twidale, 1967; Mackin, 1970 - see also Figs. 1.14 and 7.6). Three types of pediment have been recognised: those with a cover of allochthonous debris, those with a mantle derived primarily from the weathering of bedrock *in situ*, and rock pediments or platforms which are essentially devoid of a cover of unconsolidated material (Twidale, 1981c). Of these, the first type is essentially restricted to sedimentary terrains and only mantled and rock pediments are developed on granitic rock.

(a) Mantled pediments: The basic unit of the *mantled pediment* is the low-angle, fan-shaped, cone segment cut in bedrock, but carrying a thin veneer of sand, partly originating in the backing upland and transported either by small

Fig. 7.6. The piedmont angle developed at the base of a granitic inselberg near Fort Trinquet, Mauritania (R.F. Peel).

streams (where the cone occurs at the mouth of a valley) or by wash (where it is located at the base of a bluff), but mostly derived from the weathering of the bedrock of the piedmont. Some such part-cones stand in isolation (Fig. 7.7), others have merged with similar features to form either low-angle cones that surround isolated residuals (Figs. 1.12, 1.13, 1.25 and 7.8) or aprons such as those that front inselberg ranges in many parts of Australia, the Mojave region of California, and Namibia and Namaqualand in southern Africa. Such coalesced pediments are the 'pan fans' of such workers as Lawson (1915), though the use of the term 'fan' can be misleading, since it is widely used of depositional forms.

The Cima Dome and the adjacent Cimacito Dome (Fig. 7.9) in southern California are, despite their names, low-angle cones cut mainly in granite (Sharp, 1954). A similar form is developed west of Usakos in central Namibia; and there are many examples of low-angle cones surmounted by inselbergs (Fig. 1.25).

The pediments are cut in granite, but the intrinsically fresh bedrock is masked or mantled by a veneer that consists largely of weathered material *in situ*. In places, as, for example, bordering the channels of ephemeral streams, the bedrock is masked by a thin layer of stratified alluvium, but the mantle consists overwhelmingly of grus (see Twidale, 1955, 1978c, 1981c; Whitaker, 1978, 1979). Such pediments have been described from areas of pronounced winter rains, such as the Mediterranean region (see e.g. Dresch, 1949; Joly, 1949, 1952; Birot and Joly, 1952; Raynal and Nonn, 1968) and Eyre Peninsula; in

192

Fig. 7.7 (a). Pediment 'fan' or partial cone developed on granitic rocks in the Kamiesberge of central Namaqualand.

Fig. 7.7 (b). Pediment 'fans' in the Kamiesberge of central Namaqualand.

areas of continental and monsoonal climate such as Korea and in insular Japan (Akagi, 1972, 1974), and in areas that experience only episodic rains - desert and semidesert regions like the Sahara, the American West and Southwest, central Australia, and so on (see e.g. McGee, 1897; Mensching, 1958).

The plains fringing the tors of Dartmoor (Fig. 5.45) appear to be of similar type (Brunsden, 1964), though it would presumably still be considered heretical to refer to them as pediments, despite well developed covered pediments

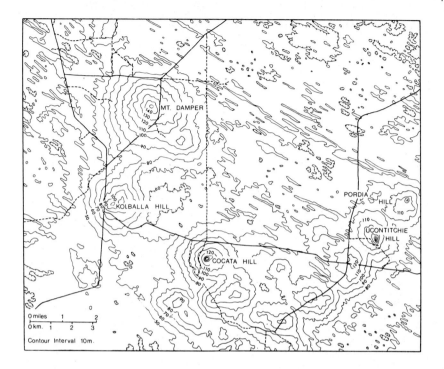

Fig. 7.8. Extract of topographic map showing part of northwestern Eyre Peninsula, S.A. (from S.A. Lands Dept. Topographic Map Series 1:50000 Pordia, 5931-IV).

occurring in such seasonally cold regions as Alaska, Utah and Colorado (Scott, 1963; Mackin, 1937, 1970; Wahrhaftig, 1970a and b; Ackerman, 1974).

(b) Rock pediments: *Rock pediments* or platforms (Twidale, 1978a) are planate rock surfaces. McGee (1897) recorded his amazement on encountering these forms in Arizona late last century. They were different from anything previously recorded:

> At first sight the Sonoran district appears to be one of half-buried mountains, with broad alluvial plains rising far up their flanks, and so strong is this impression on one fresh from humid lands that he finds it difficult to trust his senses when he perceives that much of the valley-plain area is not alluvium but planed rock similar or identical with that constituting the mountains ...

> During the first expedition ... it was noted with surprise that the horse-shoes beat on planed granite or schist or other hard rocks in traversing plains 3 or 5 miles from mountains rising sharply from the same plains without intervening foothills ... (McGee, 1897, pp. 90-91)

Rock pediments are gently inclined and are typically dimpled and grooved. Many carry remnants of a regolith, some of them with contained blocks and boulders. Some fringe uplands, others stand in isolation on hill crests, valley side slopes and in valley floors.

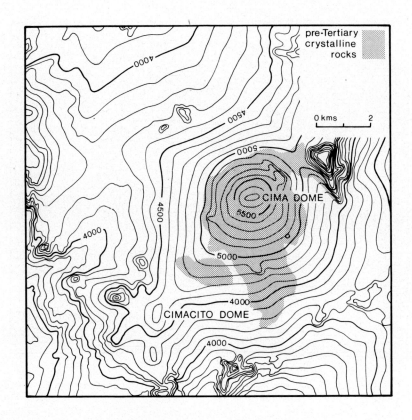

Fig. 7.9. Plan of the Cima Dome, southern California (after Sharp, 1954).

The Corrobinnie platform (see Figs. 1.22 and 7.10; also Bourne *et al.*, 1974) has already been referred to. It extends 700-800 m from the base of the associated inselberg, is scored by shallow, saucer-shaped depressions and gutters, and carries a few small boulders and patches of grus. Another extensive granite platform is developed near Tamanrasset in southern Algeria (R.L. Folk, pers. comm.) and many granitic residuals are bordered by narrow platforms, ranging in width between 1-2 m to 10 m (Fig. 7.11). Most pediments found in granitic terrains are partly mantled, partly rocky (McGee, 1897, p.91). Around Pildappa Rock, on northwestern Eyre Peninsula, for instance, there are narrow, fringing platforms (Fig. 7.12 and 7.13) and there are a few low platforms on the surrounding low-angle cone, but most of it is mantled (Fig. 7.13). The Ucontitchie pediment is similar. The surface of the mantled pediment north of Waulkinna Hill is broken by several groups of flared boulders with minor platforms exposed around them (Fig. 7.14). The bedrock surface bordering the more southerly of the Lightburn Rocks extends at least 20 m from the base of the residual, covered by only 10-15 cm of sand and exposed at several sites, and a platform related to the northern residual extends beneath a veneer of sand several kilometres to

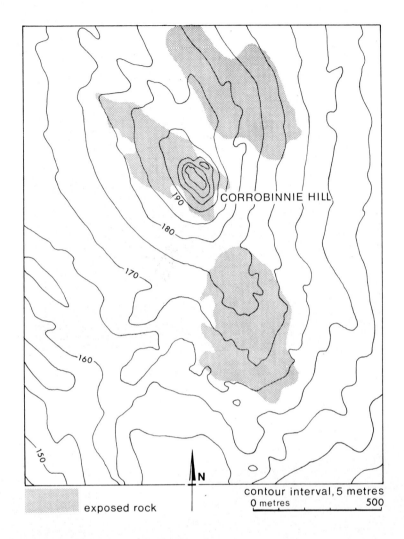

CORROBINNIE HILL

contour interval, 5 metres
0 metres   500

exposed rock

Fig. 7.10. Plan of the Corrobinnie Platform.

the east and southeast, though it is exposed, surfaces in numerous low platforms
(Fig. 7.15).

Such situations are not peculiar to granite. The plains to the west and
northwest of Ayers Rock are underlain by 3-4 m of sand (Fig. 7.16), that in the
scarp foot zone being in large measure derived from the adjacent upland but that
beyond the piedmont being mainly weathered arkose *in situ*, with only a superficial veneer of wind-blown and alluvial material. The underlying rock platform
is exposed on the northern side of the rock and at a site some 800 m west of the
residual, but the rest of the pediment is mantled (Fig. 7.17, see also Twidale,
1978b). A similar, though more exposed bedrock plain cut in conglomerate has

Fig. 7.11. Part of the southern slope of Pildappa Rock, northwestern Eyre Peninsula, S.A., showing narrow basal platform.

Fig. 7.12. Platform exposed near the southwestern margin of Pildappa Rock, northwestern Eyre Peninsula, S.A.

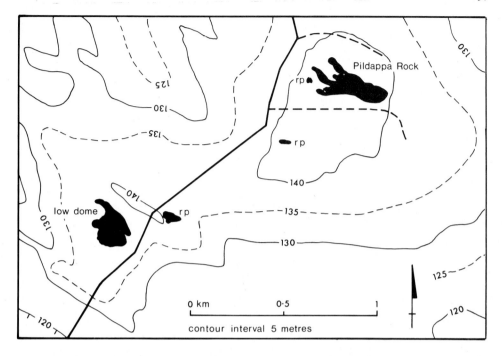

Fig. 7.13. Map of Pildappa Rock and environs, northwestern Eyre Peninsula, S.A.
r.p. - rock platform.

Fig. 7.14. Flared boulders and associated minor platforms standing on the
Waulkinna pediment (Fig. 7.5).

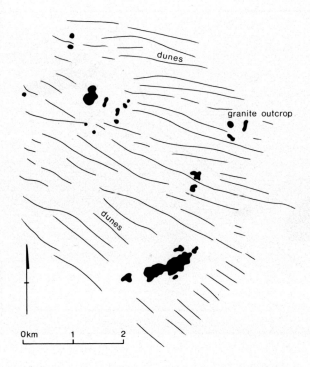

Fig. 7.15. Map of Lightburn Rocks, eastern Great Victorian Desert, S.A.

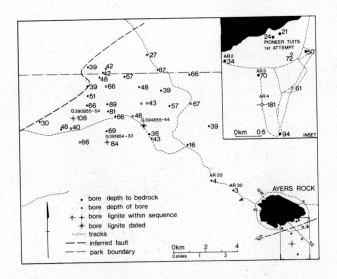

Fig. 7.16. Map of Ayers Rock and environs showing locations of bores and thickness of regolith.

Fig. 7.17. Rock platform west of Ayers Rock, N.T.

been noted as extending several kilometres west and northwest of the Olgas complex of dunes, also in central Australia.

(c) Origins: For many writers, pediments developed on crystalline rocks are, in broad view, a consequence of scarp recession (see Lawson, 1915; Bryan, 1922). Thus Howard (1942, p. 134), who worked in crystalline terrains in the American Southwest, wrote that 'the development of the pediment depends on the recession of the base of the slope ....', and Pugh, who investigated granite landforms in Nigeria, concluded that '... a mountain mass with a well-developed upper surface will shrink slowly by scarp retreat, with the development of bounding pediments.' (Pugh, 1956, p. 28). But this view is open to serious questioning for, regardless of the precise nature of the formative processes, there is compelling evidence to suggest that though in many cases (see also Chapter 5) backwearing related to scarp-foot weathering has taken place, it amounts at most to only a few tens of metres (Twidale and Bourne, 1975b; Twidale, 1978b; Twidale, 1982). The residual boulders standing on some pediments are flared, suggesting that they were marginally weathered during a period of topographic stability when the weathering front was lowered. Then followed a phase of erosion, during which the friable grus was evacuated to expose the boulders shaped by subsurface moisture attack. Thus on the Waulkinna pediment, northern Eyre Peninsula, the evidence (Fig. 7.14) suggests that there has been a recent lowering of the surface by some 2 m. Similar evidence is preserved on other, nearby, pediments, both on granite and on the dacites of the southern Gawler Ranges. The concentration of moisture in the scarp foot zone has led to the development of moats of especially deep weathering - incipient scarp-foot depressions - (Clayton, 1956;

Pugh, 1956; Peel, 1960) as well as flared forms and cliff foot caves or tafoni
(Twidale, 1962; Twidale and Bourne, 1975b), all of which argue relative stability of marginal slopes and weathering zones (see also Chapters 9 and 10).

In addition to this evidence suggestive of the stability of bounding scarps,
many of the inselbergs surmounting or backing the pediments are of limited areal
extent. Many are massive, others are well-fractured and pervious, yet all are
fringed by pediments, despite their generating little overland flow. The regolith consists largely of grus *in situ*. For these reasons any suggestion that
the pediments adjacent to granitic inselbergs are comparable to those described
from some sedimentary terrains and are due to lateral corrasion by divaricating
streams (e.g. Blackwelder, 1931; Twidale, 1978c, 1979, 1981c) must be rejected.

The field evidence suggests that mantled and rock pediments are end members
of a continuum. The mantled forms carry a veneer of weathered rock smoothed by
wash and rill work. The rock pediments are exposed areas of the weathering
front and are due to mantle-controlled planation (Mabbutt, 1966). They are the
stripped, etch, or exhumed (*sic*) forms of such writers as Paige (1912), Lawson
(1915), Davis (1938) and Tuan (1959).

Evidence for this conclusion derives from the nature of the mantle exposed on
and adjacent to the platforms; from the corestones still *in situ* and embedded
in the grus at some sites and scattered over the pediments as boulders in others;
from the physical continuity of the platforms and such features as flared slopes
that are demonstrably a particular form of weathering front (Twidale, 1962);
and from the pitted, grooved and dimpled morphology of the platforms, these minor
features being initiated at the weathering front beneath the regolith (Logan,
1951; Twidale, 1971a, p. 90, 1976b, pp. 203-204; Boyé and Fritsch, 1973;
Twidale and Bourne, 1975a, 1976a).

The mantled and rock forms are genetically related. Mantled pediments are
surfaces of transportation, as suggested by Bryan (1922), but in contrast with
the pediments of sedimentary regions the mantle of debris developed on granitic
pediments is discontinuous: running water in the form of wash and rills plus a
few minor streams has planed the surface of the regolith to give the smooth
mantled surfaces, and has also scoured the mantle to expose the bedrock in rock
platforms.

The mantled pediments are rock pediments with the regolithic cover still
intact. The regolith is important in that its contained moisture permits the
continued advance of the weathering front. But mantled pediments are not the
result of mantle-controlled planation. Rock pediments are, but mantled pediments
are not. The latter owe their surficial smoothness to wash and rill work arising either on the pediment surface or in the uplands beyond.

The relative rate of lowering of the weathering front and the mantle surface
determines the thickness of the regolith. Where the former has outpaced the

latter, the mantle is thick, but where the converse has occurred, the front is exposed as a rock pediment.

Though there are few exceptions, pediments are essentially of limited extent. Though some rock pediments occur in isolation, most granite pediments are fringing forms and extend at most only a few kilometres from the mountain front.

Pediments are particularly well developed on granitic rocks. They also show a preference for granitic environments. There are several possible reasons for this correlation between landform and rock type. First, pediments are well developed in weak rocks, and granite is particularly vulnerable to moisture attack; in any granitic massif located at or near the land surface there are characteristically large compartments of weathered material vulnerable to planation. Second, granite typically weathers to sand or grus, which is readily transported by rills and streams and spread evenly to give the smooth surface characteristic of pediments. Third, because of the low permeability of fresh granite, the weathering front is usually sharp, so that distinctive platforms or rock pediments are readily initiated and exposed. Finally, because of the marked contrast between massive and well-fractured compartments and because the latter are defined by fractures, the piedmont angle is better developed on granitic than in other lithological environments (Twidale, 1967).

(ii) Peneplains

A *peneplain* is a rolling or undulating surface of low relief and of regional extent formed by weathering, wash and river work. The broad interfluves are gently convex-upward in form. According to Davis, slopes are lowered once streams have achieved their initial major incision and when maturity has been reached and passed. Divides are lowered more rapidly than the streams are incised, and a surface of low relief is developed. Though principally of erosional origin, flood plains due to deposition extend along the main drainage lines (see Davis, 1909; see also Crickmay, 1933).

Davis himself regarded peneplains as products of what he called normal erosion, that is, river work in temperate humid conditions. In reality, however, peneplains and rolling plains that can be regarded as local, small or partial peneplains are widely distributed.

Davis cited parts of the western Great Plains in Montana and Siberia as examples, but most of the peneplains he and later workers recognised were, in fact, palaeoplains preserved on upland crests. Yet peneplain forms approximating morphologically to the theoretical model are developed on granitic rocks in the southwest of Western Australia (Fig. 1.27) and on central and northern Eyre Peninsula (Fig. 7.18). Similar features are found also in southern Africa, for instance, in northern and central Transvaal, in the western Cape Province, and in parts of central Namibia. It is true that some of these features differ in detail from the theoretical model described by Davis. In many of the areas cited streams are intermittent or seasonal in their flow, and are braided in pattern. The peneplains of northern and central Eyre Peninsula lack surface streams, save during and immediately following rains, thanks to a calcrete carapace developed during the later Pleistocene. But in general terms the examples cited conform to Davis' model. Again, the residuals that stand above peneplains are not everywhere Davisian monadnocks that rise gently from the surrounding planate surfaces; on the contrary, in many places (Fig. 7.19) insel-bergs and peneplains are frequently associated.

These peneplains are underlain partly by fresh rock, though mostly by weathered materials. This is surely to be expected, for the erosional surfaces transect various types of rock which vary in their susceptibility to weathering. Moreover, in many instances it is not clear whether the surface is cut in pre-viously weathered bedrock or whether the alteration is related to the same phase of development as the planation. Whether fresh rock or regolithic material is exposed in a surface of low relief is not diagnostic (cf. Dury and Langford-Smith, 1964).

Peneplains are not restricted to temperate lands. Indeed, they are not con-fined to any conventionally defined climatic region. Like pediments, they are characteristically developed on weak rocks, typically argillaceous sediments or weathered crystalline rocks.

It is difficult to determine whether slope decline or backwearing has been involved in their development. On general grounds it can be argued that in areas of weak rocks downwearing ought to have been dominant, but there may have been local or ephemeral complications, such as aridity and the development of a gibber veneer, or the development of a duricrust, in either case implying a caprock and scarp retreat. The end result was the lowering of the land surface, but there is no means of demonstrating past slope behaviour. Moreover, slope morphology may reflect only the most recent retouching of the land surface and not the influence of other processes that have been operative for most of the period during which the surface was evolved.

(iii) Relationship between peneplain and pediment

Rock platforms are an integral part of some peneplains. Thus, on Eyre Peninsula they occur on the crests of low, convex rises, in valley floors and in gentle hillslopes. Whether such platforms constitute the upper margins of just-exposed massive compartments, or are the last remnants of old residuals that have been completely reduced is an interesting question, but one that is not only irrelevant in the present context, but non-susceptible of solution.

Again, the fringing mantled pediments that are developed around such insel-bergs as Ucontitchie Hill merge at their lower extremities with the rolling surface of the peneplain, and without topographic break (Fig. 7.8). Both forms are developing concurrently. Moreover, there are enough boulder clusters and

Fig. 7.18. Parts of the Eyre Peninsula (S.A.) granite peneplain seen (a), above, between Minnipa and Wudinna, in the northwestern region, and (b) below, in the northeast, between Iron Knob and Kimba.

other low residuals to suggest that when the latter are eventually eliminated the pediments that now surround them will extend and coalesce in convex crests that are integral parts of the peneplain.

Thus rock and mantled pediments are not genetically or temporally distinct from peneplains. They are particular types of plains surfaces developed on fresh rock and in the piedmont of inselbergs respectively. Pediments are not fundamentally distinct from peneplains: they are part of them. Moreover, though pediments are well and widely developed and preserved in arid and semiarid lands, they occur also in other climatic contexts. Pediments are typical of the arid and semiarid tropics but they are not, as Blackwelder (1931, p. 138) suggested, merely 'the desert-inhabiting species of the genus peneplain': both pediments

Fig. 7.19. Inselberg and peneplain in granitic rocks, Vredenburg areas, western Cape Province, R.S.A.

and peneplains are developed on granitic rocks in varied climatic contexts.

(iv) Pediplain and Ultiplain

Apart from peneplains, there are, in various parts of the world, remarkably flat, featureless and extensive plains, some of which are cut in granite rocks. Several of them have been called *pediplains* which King (1942) saw as the extreme stage of cycles of pediplanation:

> During maturity of the landscape opposing scarps meet from the opposite sides of hills, which are thereby rapidly lowered. The residual uplands disappear, relief decreases markedly and with coalescing of the ever-increasing pediments a bevelled landscape of low relief and of *multiconcave form, a pediplain* ... is produced. (King, 1942, p. 53).

King cites the Springbok Flats, cut in basaltic rocks, as an example of such an extreme stage of pediplanation. Many such planation surfaces, however, are really very low relief peneplains, because there is some relief, and the broad divides are convex. Even the Springbok Flats display some relief.

Some so-called pediplains are alleged to be multiconcave surfaces. Certainly many foot-slope pediments are concave-upwards, though a few are convex. But no extensive planate surface with angular divides has been described.

Nevertheless, planation surfaces of nil relief are developed in various parts of the world. Morphologically, they differ from peneplains and warrant separate consideration.

The Bushman (or Bushmanland) Surface of the northern Cape Province is a plain of regional extent and extreme flatness that is eroded in granite, gneiss and sandstone (Fig. 1.14), and the plains around Meekatharra, in central Western Australia, eroded in Archaean migmatites, are similar (Fig. 1.28).

Morphologically these plains are comparable only to the depositional plains of central Australia (Fig. 7.20) and elsewhere. Yet they are manifestly of erosional origin, for intrinsically fresh bedrock occurs only a few centimetres beneath the surface. The Bushman Surface near Gamoep, in central Namaqualand, is of quite remarkable flatness, and even low residuals of very minor areal extent stand out as prominent features of the landscape (Fig. 1.14). Yet, wherever there has been shallow gullying or shallow excavation for road metal or to create water storage, fresh granite or other country rock is exposed less than a metre beneath the land surface.

The very flat plains have commonly been considered to be the end product of scarp retreat and pedimentation, and to result from the coalescence of pediments (e.g. King, 1942, p. 53, 1950, 1953, 1962; Dresch, 1957; Pallister, 1960). But this suggestion is difficult to sustain as a general argument. It has been shown that scarps retreat only in certain structural circumstances (Twidale, 1978c). Pediments are not necessarily associated with backwearing, for some scarps are essentially fixed in space and yet pediments are associated with them (Twidale, 1978c). Pediments are restricted to the scarp foot zone where coarse, or relatively coarse, debris has been deposited as a protective veneer. The only exceptions are rock pediments which are confined to resistant outcrops and which are, in any case, of limited extent and relatively rare.

There appears to be no necessary or even likely genetic connection between pediments and pediplains. The former are a particular type of footslope which,

Fig. 7.20. Flat alluvial plains adjacent to the Georgina River, near Bedourie, southwest Queensland.

though widely distributed, are yet restricted to the piedmont and extend only
a few kilometres, at the most, from the scarp foot. The latter are simply
peneplains that have been baselevelled to the ultimate degree.

These flat plains are merely peneplains that have been further degraded.
They represent a further stage of denudation, as Peel (1941) clearly appreciated.
They are most likely to evolve in areas of tectonic stability and to be preserved
in areas of arid climate. The examples cited are located in arid and semiarid
areas of Africa and Australia and in shield areas that have suffered little
tectonic disturbance since the middle Proterozoic or even earlier, for over wide
areas of both of these continents Upper and even Middle, Proterozoic strata
remain essentially undisturbed.

Streams, rills and wash have so degraded the land mass until there is no
relief. Aridity implies slow rates of landscape change (Corbel, 1959) and, in
particular, a virtual absence of stream activity for extended periods. In
addition, coarse debris (gravel - hamada, reg, serir, gibber, desert armour,
desert pavement) tends to accumulate and both smooth and protect the surface
(see also below).

Of course, there are variations and complications. Where, as in parts of the
northern and central Transvaal, streams have recently been rejuvenated, the
flat plains are fairly dissected to give a peneplain form (Birot *et al.*, 1974).
Elsewhere, plains have been stabilised by the formation of duricrusts. Thus
calcrete has accumulated at and near the surface of much of the peneplain that
occupies central, northern and northwestern Eyre Peninsula (Twidale *et al.*,
1974, 1976) and, until the calcareous carapace is destroyed, the rolling topography will be preserved.

Support for the suggested genetic relationship between peneplains and the
very flat surfaces derives also from a consideration of their ages. Several
extensive peneplains are associated with present baselevel; they are, in geological terms, youthful features and most of them are of later Cainozoic age.
The very flat surfaces, on the other hand, seem to be much older. The Bushman
Surface, for example, has a long and complex history, but is regarded as being
of at least Early Tertiary age (King, 1962, p. 266; see also Mabbutt, 1955).
In all probability, however, it is older and may, in part, be exhumed and of
Palaeozoic age. The Meekatharra Surface (Fig. 1.24) appears to be of the same
age as the Hamersley Surface to the north, that is, probably Cretaceous (R.C.
Horwitz, pers. comm.), for it stands higher than valley floor remnants that are
capped by various duricrusts, including some of probable early Cainozoic age,
and for this reason alone is tentatively regarded as later Mesozoic.

Thus, it is suggested that the plains of virtually no relief have no genetic
connection with pediments save insofar as the latter are a special type of
valley-side slope that may eventually be degraded to become part of a low relief

surface. The very flat plains are simply plains of low relief of great antiquity. They have been stable and exposed for so long that wash, rills and rivers
have gradually smoothed and pared away even the low swells of peneplains. If
peneplains are taken literally to be penultimate plains or 'almost plains', then,
as Davis remarked, 'the ultimate stage would be a plain without relief' (Davis,
1909, p. 270). The flat, featureless plains must, surely, be these ultimate
plains. It would be misleading to refer to these very flat plains as pediplains,
for both the word itself and past usage suggest a genetic connection with pediments, whereas the evolutionary relationship is with peneplains.

They are the forms for which the terms 'desert peneplain' and 'surface of
maximum denudation' have been proposed by Sandford (1933), but the former merely
suggests a peneplain in a desert situation and so does not suggest morphological
contrasts between the two planate forms, while the latter is slightly ambiguous.
Bagnold (1933) and Peel (1941) both refer to the extraordinary plains of
southern Libya and the adjacent parts of the Sudan as sand sheets, but this can
be confused with depositional forms of the desert and, in any case, not all of
the very flat plains carry a veneer of sand. Hills (1955) and Öpik (1961) proposed the term 'old-land'. Unfortunately the term is pre-empted, for it has
been used of extensive remnants of ancient crystalline rocks reduced to low
relief by long-continued erosion, so that Waters (1957, p. 503), for instance,
referring to the several discrete granite high plains of the region, writes of
the 'old land of south-west England'. Again, earlier this century such geomorphologists as Davis (1909), Lobeck (1939) and Cotton (1948, p. 89) conceived of
older areas of mature landscape bordered by newer coastal plains. Several
tectonists, notably Suess and Kober (see Steers, 1932), discussed old shield
lands. In all these circumstances it would be confusing to persist with a term
that has been used in slightly different senses, either in the same or in
approximate form. The French term *surface d'aplanissement* (or planation surface
- e.g. Baulig, 1952), though useful, is too general.

If a peneplain is almost a plain it can be suggested that these, the ultimate
plains (see Davis, 1909, p. 270, in the phrase cited above), which can become no
flatter, be called *ultiplains*, the term being used to imply merely plains of nil
relief resulting from very long-continued degradation, but without reference to
mode of development, without implication as to downwearing or backwearing of
slopes.

(v) Multicyclic and stepped assemblages

Multicyclic forms are developed in granitic terrain as a result of the
relative lowering of baselevel, stream rejuvenation and landscape revival. The
plains become high plains, located high in the relief, as in the Sierra Nevada
(Fig. 5.4), in the Rocky Mountains of Colorado and Wyoming (the Sherman Surface
of Eggler *et al.*, 1969; see also Lee, 1922), on Dartmoor and several parts of

southern Africa. But they are dissected by streams working toward their new
baselevel. Valley-in-valley forms are first developed. These comprise valley
side facets separated by breaks of slope (Fig. 7.21) and corresponding graded
stream sectors separated by nick points in the shape of waterfalls or rapids.
Of course, in granitic terrains such breaks of slope may develop for structural
reasons. Thus, particularly massive blocks may form local baselevels to which
stream sectors are graded. Similarly, and especially in areas of gneissic
rocks, particularly massive bands of rock may give rise to stepped relief on the
valley side slopes, as, for instance, in the Rooifontein Valley of central
Namaqualand (Fig. 7.22) where such local baselevels can be related to structural
factors. However, a similar morphology is developed on the opposite side of the
valley, suggesting that the forms are cyclic. In like fashion, if waterfalls
are cyclic rather than structural they ought to be developed on all rivers in a
given region, whereas structurally determined features tend to be random or
isolated.

Fig. 7.21. Summit planation surface (1) and valley side faces (2 and 3),
Kamiesberge of central Namaqualand, R.S.A.

Valley side facets tend to extend laterally and nick points inland so that,
in time, a new surface of low relief comes to replace the former plain.
Remnants of the latter, and indeed of even earlier surfaces, may persist in the
landscape, so that the latter has a stepped appearance and can be described as
multicyclic since there is in it evidence of more than one geomorphic cycle (see
e.g. Fig. 7.21).

A rather different interpretation of stepped topography is due to Wahrhaftig
(1965) with respect to the southern Sierra Nevada (Fig. 7.23). Planation surfaces are a prominent feature in many parts of the upland. Three major high
plains have been recognised in some areas (Matthes, 1937) and, in addition,

Fig. 7.22. Stepped valley side slopes in gneiss, near Rooifontein, Kamiesberge, central Namaqualand, R.S.A.

many minor ones. The slopes separating adjacent surfaces have long been recognised and variously interpreted as fault scarps (Hake, 1928) or as due to
differential erosion and lithological contrast (e.g. Krauskopf, 1953), but the
field evidence sustains neither of these views.

According to Wahrhaftig the stepped topography is not only confined to
granitic outcrops but is found wherever granite is exposed in the region.
Wahrhaftig explains the forms in terms of the behaviour of granite in wet and
dry environments:

> The stepped topography is believed to be caused by differences in the rate
> of weathering in the two environments to which granitic rocks in the Sierra
> Nevada are subject. Where buried by overburden or gruss, the solid granitic
> rocks are moist most of the year, and disintegrate comparatively rapidly ...
> where exposed, the solid granitic rocks dry after each rain and therefore
> weather slowly. (Wahrhaftig, 1965, p. 1166).

Wahrhaftig points out that dry granite is stable and resistant whereas
granite in contact with water is rapidly altered (cf. Barton, 1916; Bain, 1923).
Thus blocks exposed in linear zones tend to become upstanding and to form local
baselevels for upstream sectors of rivers crossing the barriers (Fig. 7.24). In
this way a stepped topography evolves, dependent in the first instance on the
varied behaviour of granite to weathering. Wahrhaftig discusses various possible
reasons for the upstanding outcrops becoming upstanding in the first instance,
but fracture density is almost certainly the cause - the ridges are based on
outcrops of massive blocks.

An alternative explanation of stepped topography can be suggested, using the
scarp-foot weathering responsible for stepped inselbergs (see Chapter 5;

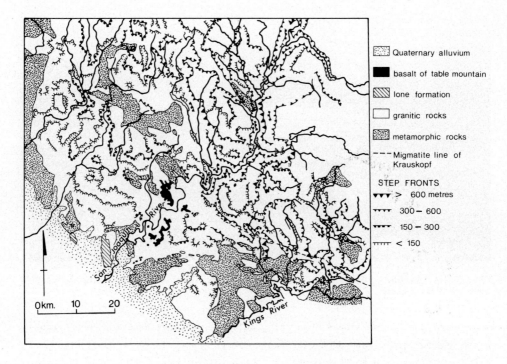

Legend:
Quaternary alluvium
basalt of table mountain
Ione formation
granitic rocks
metamorphic rocks
--- Migmatite line of Krauskopf

STEP FRONTS
▼▼▼ > 600 metres
▼▼▼ 300 – 600
▼▼▼ 150 – 300
▼▼▼ < 150

Fig. 7.23. Stepped topography, southern Sierra Nevada, California (after Wahrhaftig, 1965).

Fig. 7.25). Given an initial, structurally determined break of slope, scarp-foot concentration of moisture causes alteration of the rock, development of a steeply inclined weathering front (typically in the form of a concave-upward flare) which is worn back as a result of continued moisture attack. The steep subsurface cliff extends laterally in a gently sloping sector of the weathering front which also develops a shallow concavity as the recession of the inner front

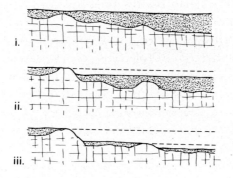

Fig. 7.24. Development of stepped topography according to Wahrhaftig (1965).

proceeds. Water from the residual flows to the lower margins of the rock and
some of it percolates into the regolith and to the weathering front. Most of it
is concentrated in the scarp foot zone, so that the depth of weathering is
greater near the margin of the exposed rock than it is a few metres away. Thus,
as the weathering front recedes, a deeper moat of weathered debris comes to be
formed, framed at its outer edge by a slightly higher rim of fresh rock.

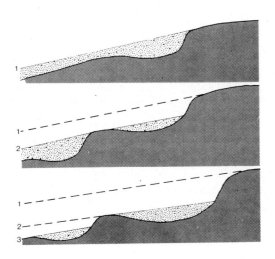

Fig. 7.25. Development of stepped slopes by repeated alternations of scarp
foot weathering and baselevel lowering.

Lowering of baselevel by an amount equal to or greater than the depth of the
regolith causes the weathered rock to be eroded and the weathering front, with
its steep inner margin, broad, shallowly concave-upward platform and outer rim,
to be exposed. Scarp foot weathering again develops at the outer edge and,
again, the steep inner face of the weathering front causes recession of the
edge of the fresh rock outcrop. Repetitions of phases of weathering and erosion
cause stepped topography to evolve (Fig. 7.25 and 7.26). Grus and soil persist
on some platforms in the shallow depressions between steep inner scarp and the
outer rim, but elsewhere this debris has been evacuated by streams draining
laterally, parallel with the topographic fronts, into fracture-controlled
clefts.

But whatever their origin these limited granite plains gradually extend
until the whole region is baselevelled.

Granite plains of various types are developed in several parts of the
world, though they are best preserved in lower latitudes. Their widespread

Fig. 7.26. Sketch of the stepped slope of Poondana (Brazil) Rock, northwestern Eyre Peninsula, with soil-filled depressions occupying the inner areas of the treads; s - soil, r - rock; the numbers indicate the sequence of planate forms developed.

and extensive development reflects, first, the vulnerability of most granite masses to weathering, and particularly to moisture attack, and, second, the virtual ubiquity of streams and rivers.

PART III

MINOR LANDFORMS

CHAPTER 8

FORMS OF GENTLE SLOPES

Some bornhardts have gently sloping upper surfaces. Like platforms, which in many instances they were before the plains around them were lowered, they are commonly dimpled, due to the development of rock basins, and grooved by clefts and runnels (Fig. 8.1). In addition, though more rarely, pedestals and doughnuts are developed at some sites.

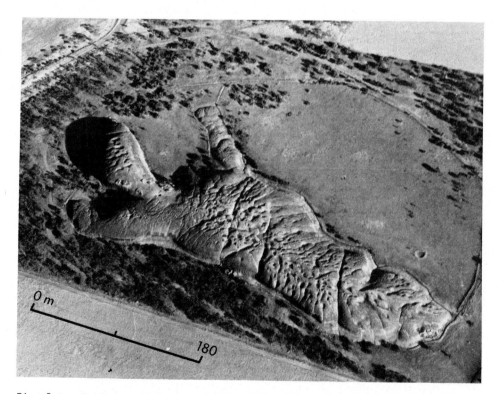

Fig. 8.1. Oblique aerial view of Pildappa Rock, northwestern Eyre Peninsula, showing prominent NW-SE vertical fractures. The upper surface is dimpled due to the development of rock basins, and the southern slope is flared (see Fig. 7.11).

A.   ROCK BASINS

(i) Description

Rock basins are depressions excavated in solid bedrock. Morphologically they
vary in detail, but most are oval, elliptical or cicular in plan. Some, strongly influenced by jointing (Figs. 8.2 and 8.3), are angular in form, but others,
resulting from the coalescence of two or more individuals, are irregular in plan.
Several morphological types have been recognised (Fig. 8.4):

- *pits* are hemispherical and developed on gentle slopes;

- *pans* are comparatively shallow, are flat floored, and are also developed on
gentle slopes;

- *cylindrical hollows* vary in plan shape, though they are generally circular
and are rectangular in vertical section so that they are appropriately referred
to as being of cylindrical form;

- *armchair-shaped hollows* are asymmetrical in section normal to the contours,
having high backwalls on the upslope side, but being open downslope. They are
typical of the moderately steep (20°-30°) slopes that lead down from the flattish crestal areas to the steep bounding slopes and the plains (Fig. 8.5).

Rock basins of various types have been described from many parts of the world
and from many different lithological and climatic settings. Pans are particularly common and widely distributed. They are as well developed in polar as in
equatorial lands, in deserts as in humid regions. They have been described in
granitic rocks from areas as climatically diverse as the Yosemite region of the
Sierra Nevada of California (Matthes, 1930), and the Karkonosze Mountains
(Jahn, 1974); Mongolia (Dzulinski and Kotarba, 1979); the humid subtropics of
Georgia, in the eastern U.S.A. (Smith, 1941); the Snowy Mountains of N.S.W.;
Surinam and Brazil; and so on - they are practically ubiquitous. They are
developed in sandstone in the southern Flinders Ranges, South Australia; in
several parts of Europe (cf Klaer, 1957); in Tasmania; and in the Drakensberg
of South Africa (Twidale, 1980a); in arkose on Ayers Rock, central Australia;
in the dacite of the Gawler Ranges, South Australia; and widely, of course, in
limestone.

(ii) Nomenclature

These various depressions have been referred to by several names through the
years in various parts of the world. Though first mentioned in the scientific
literature in the late Eighteenth Century, the forms had actually been perceived
and discussed in general terms for many centuries (see Worth, 1953, p. 30).
They were, as far as can be ascertained, first described from Dartmoor in southwestern England, where they were known as rock basons (Borlase, 1754) or rock
basins (MacCulloch, 1814; Ormerod, 1859; Worth, 1953).

Rock basins of various types have been referred to also as rock holes,
weather pits (see e.g. Smith, 1941), pot holes, water eyes (see e.g. Tschang,

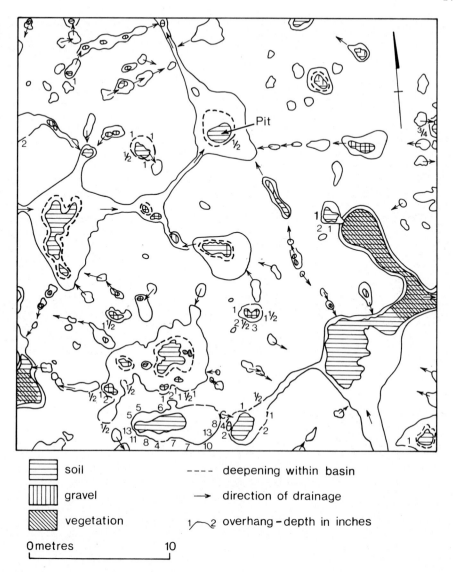

Fig. 8.2. Plan of part of upper surface of Pildappa Rock, northwestern Eyre Peninsula, S.A., showing angular basins influenced by jointing.

1962), cauldrons (see Ormerod, 1859), granite pits, and, in Australia, night-wells (Maclaren, 1912), *gnammas* (Carnegie, 1898; Maclaren, 1912; Jutson, 1914; Twidale and Corbin, 1963), or alternatively as gnamma holes, which, however, is tautological as, in many Aboriginal languages, gnamma means hole or rock basin. Anderson (1931) refers to Idaho examples as bath tubs, and other local names have doubtless also been applied. These forms are the *tanques* and *vasques rocheuses* of the French literature, and the *Verwitterungsnäpfe, Opferkessel,*

Fig. 8.3. Pan, shape in plan influenced by fractures, near Johannesburg, R.S.A.

Fig. 8.4. Various forms of rock basins: (a) pit, on Pildappa Rock.

Fig. 8.4 (b). Pan with overhanging sidewalls on the Kulgera Hills, central Australia.

Baumverfallspingen, *Felsschüssel* and *Dellen* of German workers. They have also been called *kociolki* in Polish, *pias* and *pilancones* in Spain, *caldeiros, pocos, marmitas* and *oricangas* in Brazil, and *araceenhorst* in Surinam (see Bakker, 1958).

(iii) Origin

(a) General comments: Rock basins were, for many years, ascribed to human agencies (Borlase, 1754; Drake, 1859), and in particular were thought to be related to Druidical ceremony. In ancient Germany, too, they were regarded with awe as their name - *Opferkessel*, or sacrificial cauldron - suggests. The red staining due to the release of iron by the weathering of biotite, or due to red algal growths in the pools of water, suggested blood to the ancients. The arguments adduced in support of this explanation now make interesting reading. Drake (1859, p. 370), for instance, asserted that the basins are too perfectly rounded to be other than man-made, and, challenging the idea of a natural origin reiterated at that time by Jones (1859) and Ormerod (1859), wrote 'Surely, if the atmosphere is the cause, these rock basins might be found in all granite regions and in all latitudes'. But, as one of the earliest scientists to con-sider rock basins (MacCulloch, 1814, p. 43) commented, the numbers of rock basins developed on granite outcrops are such that the Druidical hypothesis 'Must ... have required a priesthood sufficient to exclude all other population, if every rounded cavity which the granite exhibits was a pool of lustration',

Fig. 8.4 (c). Armchair-shaped hollows on Pildappa Rock.

i.e., of purification by such rites as ceremonial washing or expiatory
sacrifice.

There is ample evidence that basins are due to moisture attack, which produces rounded forms of considerable perfection through the concentration of
weathering on projections.

Several early workers (MacCulloch, 1814; Ormerod, 1859; Jones, 1859) noted
that granite sand remains in the floors of many of the basins, that the mica and
feldspar are altered by moisture, and relating these two observations, concluded
that water is capable of causing rock disintegration: as MacCulloch (1914,
p. 73) put it 'We need not hesitate in admitting the solution of granite in
water to an extent capable of producing the effect of disintegration ...'. On
the other hand, it seems clear that in cold lands frost shattering contributes
to the formation of the basins (see e.g. R. Dahl, 1966; Sugden and Watts, 1977).

It is difficult to ascertain with certainty why specific basins are located
where they are, for the rock that provides the evidence has been weathered and
evacuated. Some, but not all, have formed along or at the intersection of
fractures and fissures. It would be mistaken, however, to assume that granite
is a homogeneous rock (see Chapter 2). Xenoliths are fairly commonplace and the

Fig. 8.4 (d). Cylindrical hollow on Kwaterski Rocks, northwestern Eyre
Peninsula, S.A. Diameter 0.8-0.9 m.

essential minerals of granite are not everywhere evenly distributed. It is not
uncommon to find bands or discrete masses of feldspar or biotite, both of which
minerals, being susceptible to moisture attack, could be preferentially weathered
to form an initial hollow that could later develop into a basin.

But, having indicated other possible origins, it must be reiterated that
water is primarily responsible for the initiation and growth of rock basins.
Quite apart from the evidence and argument already adduced, additional proof of
the natural, as opposed to any form of anthropogenic, origin of rock basins were
needed, it is provided by the observation that incipient saucer-shaped depressions are developed at the weathering front beneath the regolith (Fig. 8.6).
Both in the subsurface and on exposures, fractures are exploited and basins
developed along them (Figs. 8.7 and 8.8).

(b) Rate of development: The evidence concerning the rate of growth is
equivocal. Rock basins have developed on scree blocks in some areas (Twidale
and Corbin, 1963) suggesting that the forms evolve quite rapidly in geological
terms - in a matter of a few thousands of years. Basins are widely and well
developed in several polar regions (e.g. Scandinavia, the Canadian Arctic) that
were covered by ice until a few thousands of years ago. The implication is that

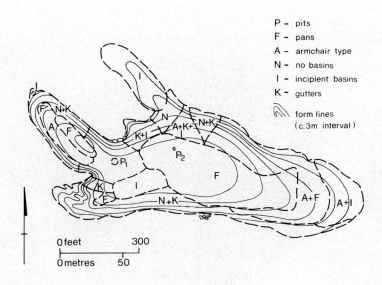

P - pits
F - pans
A - armchair type
N - no basins
I - incipient basins
K - gutters

form lines
(c.3m interval)

0 feet   300
0 metres   50

Fig. 8.5. Distribution of rock basins on Pildappa Rock, northwestern Eyre
Peninsula.

the basins have either survived glaciation or have developed since the exposure
of the surfaces on which they occur. The latter is not impossible, for in the
Bohemian Massif, Czudek *et al*. (1964) report basins developing at a rate of
several centimetres per century (see also Chapter 3), and, though the measurements were not as precisely controlled as might be thought desirable, there is,
to say the least, strong suggestion of rapid growth.

Some workers (Reid *et al*., 1912, p. 73) suggested that certain basins in the
southwest of England are deepening at a rate of 0.25 cm per annum, but this is
not supported by other observations. As Worth (1953, p. 30) has pointed out,
Mistor Pan, which is a large and well-known rock basin on Dartmoor, was described
as early as 1291 and again in 1609. Its depth was measured in 1828, 1858, 1875
and 1929. All results were similar, the differences being as readily accountable
in terms of the difficulty in determining the upper limit of the depression as
by any real increase in depth.

It may be that some rock basins are no longer actively forming, though the
accumulation of sand and other detritus does not constitute evidence of stability.
On the contrary, such accumulations retain moisture that is responsible for
extending the depressions.

On the other hand, a decrease in rate of growth may be an integral feature of
basin development. Any given hollow drains a certain area of the bedrock surface and, over the years, a certain average volume will be contributed to the

Fig. 8.6. These rock basins, developed along major fractures, were exposed when the regolith, consisting entirely of grus, was stripped away during quarrying operations near Ebaka, South Cameroon (M. Boyé).

basin. While the diameter of the basin is small, this volume X can be expected to fill the depression for a considerable part of each year, so that contact between water and basin walls is of long duration and growth is rapid. But as the diameter of the pool occupying the basin increases, evaporation losses increase and the volume X will not only moisten less of the surface but will do so for fewer days each year. Thus development ought theoretically to take place at a decreasing rate.

(c) Differentiation of major types: After exposure, the saucer-shaped depressions are further developed and several morphological types evolve.

If moisture attack were the only factor determining the development of rock basins then pits would be formed everywhere. Because water is in contact longer with the floors and lower sidewalls of the depressions than with the upper sidewalls, weathering would effectively cause the basin to be extended vertically

Fig. 8.7. Rock basins developed along prominent fracture zone on Peella Rock, northern Eyre Peninsula.

into the rock mass and, depending on whether or not there was undermining of the surface layer, a pit or a flask-shaped depression would be formed (Fig. 8.9). And in some areas such forms are indeed developed: water acts on sensibly homogeneous granite, and it is the duration of wetting that is all-important. Thus at Lightburn Rocks, eastern Great Victoria Desert, pans are well developed on the crestal areas of flaggy or flaky rock, but pits are found on the lower gentle slopes of the flanks where massive rock is exposed.

At some sites many small, comparatively shallow, pits have developed in close proximity to one another, so that the basins are separated by knife-edged ridges: *in toto* they look like a choppy sea, or the surface of a meringue: hence meringue surface (Fig. 8.10).

The most common type of basin, found in various climatic and lithological settings, is the flat-floored pan. Pans are circular or elliptical in plan and their outlines are generally smooth and regular, most projections having been eliminated by moisture attack. They are generally 2-3 m in diameter and some 30-50 cm deep, though composite features are, of course, larger and also less regular in plan. The sidewalls of many, perhaps most, pans are overhanging by a matter of a few centimetres, though on Yarwondutta Rock, northwestern Eyre Peninsula, the floor of one basin extends about one metre beneath the edge of the protruding lip (Fig. 8.11). Pans are consistently developed in flaggy rock

Fig. 8.8. Elongate pit formed along fracture at Lightburn Rocks, eastern Great Victoria Desert, S.A.

characterised by discontinuous fractures that run essentially parallel to the rock surface. Water readily penetrates along these joints so that lateral extension takes place more rapidly than does vertical growth. This explains the large diameter/depth ratio of the pans. The floors of the pans are frequently coincident with parting planes (Fig. 8.9).

The overhanging sidewalls of the pans are due to two effects, one structural, the other related to duration of wetting. First, in many areas, for example on the Kulgera Hills, a substantial superficial induration of oxides of iron and manganese is developed, and this effectively cements the bedrock so that it withstands weathering and erosional attack better even than does the fresh rock. Second, water persists longer in the floor of the depression and the lower

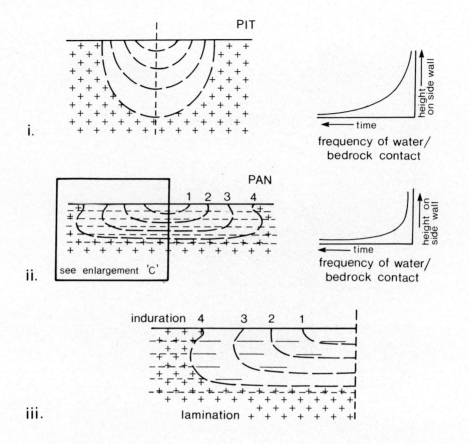

Fig. 8.9. Schematic development of pit and pan.

sidewalls are, for this reason, weathered back more rapidly than are the higher
zones. This last factor is everywhere important, for overhanging sidewalls are
developed in many areas, such as Dartmoor and the Yosemite, where there is no
detectable surface induration.

The detailed distribution of the pans and pits on such residuals as Pildappa
Rock is consistent with the suggested explanations of the two morphological
types. Pans are far and away the more numerous on the broadly rolling upper
surface, though armchair hollows are prominent on the steeper marginal zones
(Fig. 8.5). There are, however, two pits, and both occur in topographic lows,
below the level on which pans are so profusely represented. One is located on
a col between the western snout and the main body of the rock, the other in a
joint-controlled cleft within the upper surface and also within a pan developed
along the fracture zone. On the col, in particular, the presence of isolated,
blocky residuals shows that a layer or sheet of rock, up to a metre thick, has

Fig. 8.10. A meringue surface formed as a result of the development of numerous shallow pits, the margins of which intersect to give sharp-crested miniature divides, Scrubby Peak Station, northwestern Eyre Peninsula.

Fig. 8.11. This flat-floored pan on the crest of Yarwondutta Rock is almost 1 m deep and has an overhang almost as great on its northwestern margin (facing camera).

been stripped away. It is suggested that here, and in the joint cleft, it is
the upper flaggy zone that has been eroded, exposing the homogeneous rock in
which pits have evolved.

   This explanation may be of general application. Ormerod observed only pans
in southwestern England and he expressly stated (Ormerod, 1859, p. 16) that the
superficial zone of the granite outcrop, on which the rock basins were developed,
is flaggy. Likewise, Smith (1941) described only flat-floored pans, and though
he did not explicitly mention partings, his illustrations (e.g. his Fig. 4 at
p. 21) strongly suggest that the bedrock of his study area is horizontally fractured. The general point was appreciated more than a century ago by Jones (1859,
p. 311), who wrote that the tabular (or flaggy) structure of the Dartmoor granite
(see Chapter 2) was 'probably the cause of frequent occurrence of basins with
flat bottoms'. Of course, where basins develop in granite that is sensibly isotropic, save for the development of a superficial induration, flask-shaped basins
or *Opferkessel* develop.

   Like pans, pits are in some places influenced by structure. The most
dramatic results are found where the pits have penetrated through the slab or
sheet in which they have developed to the sheeting joint beneath. Water flowing
to the erstwhile pit cannot be contained, for it runs along the basal fracture
so that the previously enclosed basin becomes a throughway for running water.
The base is widened, the water swirls around, creating regular spiral patterns
and a cylindrical hollow (Fig. 8.12) is formed (Twidale and Bourne, 1978b).
Many of these hollows become choked with debris, but some remain clear, the
sheeting joints become enlarged, and hollows or tafoni are incipiently developed
beneath the sheet. Such cylinders and subsurface tafoni have been noted at two
sites on northwestern Eyre Peninsula (John K. Rock and Myrtle Rock), and at
Lightburn Rocks in the eastern Great Victoria Desert. On the western, upper
slopes of Ucontitchie Hill, also on northwestern Eyre Peninsula, there is a small

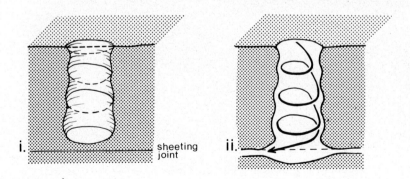

Fig. 8.12. Stages in development of a cylinder.

cylinder which, like a so-called cauldron described from Dartmoor (Worth, 1953, p. 32), cuts through two sheets of rock, with widening and incipient tafoni developed along each of the horizontal partings.

Thus the contrasts between the essential morphological characteristics of pans, pits and cylinders can be explained in terms of structure and duration of wetting. These forms develop on gentle slopes and each has its analogue in the steeper marginal areas where armchair-shaped hollows of asymmetrical cross-section predominate.

Evacuation of debris: The evacuation of debris from the basins has given rise to some discussion. It is to some extent a non-question, for much of the granite sand remains in the basins. Where, however, they are essentially clean, water flushing through the drainage network and transport in solution offer feasible mechanisms. Wind is not a significant agent. Many rock basins, both in Australia and elsewhere, have been cleared of debris. Not only did the Aboriginal population clean out the basins in order to make them more effective as water storages, but European shepherds and stockmen have done the same. Both Aboriginals and Europeans placed slabs of rock (and the latter, sheets of corrugated iron) over basins in order to reduce evaporation and contamination.

(d) Initiation at the weathering front: In an earlier paper (Twidale and Corbin, 1963) it was argued that water collecting in chance crevices and depressions in the exposed rock surfaces could enlarge those initial depressions to form rock basins. A similar view was expressed much earlier by MacCulloch (1814, p. 73) who argued that 'If a drop of water can make an effectual lodge-ment on a (granite) surface ... a small cavity is sooner or later produced. This insensibly enlarges as it becomes capable of holding more water ...'. Wilhelmy (1958, p. 37) and Schmidt-Thomé (1943, p. 55) also regarded rock basins as exogene forms and Hedges went so far as to state that 'They are never found to be developed beneath a soil cover in road cuts and other excavations' (Hedges, 1969, p. 26).

Notwithstanding such assertions, observations from many sites confirm what was appreciated in general terms some 130 years ago (Logan, 1849, 1851), namely that many minor granite forms, including rock basins, are initiated in the sub-surface (Figs. 8.6 and 8.13).

Northwestern Eyre Peninsula receives only a moderate and unreliable rainfall. Consequently, considerable attention has been devoted to water conservation and, over the past fifty years, many storage reservoirs, large and small, have been excavated at the lower margins of the granite residuals (Twidale and Smith, 1971). The weathered granite or grus has been stripped away, exposing the weathering front formed in intrinsically fresh bedrock. On exposure, the weathering front is seen to be dimpled and grooved due to the development, beneath the regolith cover, of shallow, saucer-shaped depressions and linear

Fig. 8.13 (a). These saucer-shaped depressions were exposed when weathered granite was stripped to reveal the weathering front during the excavation of a reservoir at Dumonte Rock, near Wudinna, northwestern Eyre Peninsula. X - X marks the natural soil level.

channels. Such evidence has been revealed at many sites on Eyre Peninsula (see Twidale, 1971a, p. 90; 1976b, pp. 203-204; Twidale and Bourne, 1975a) and elsewhere (e.g. Halfway, between Johannesburg and Pretoria, and at Ebaka Quarry in South Cameroon - see Boyé and Fritsch, 1973). The several types of rock basin develop after exposure, but the basic depression is initiated at the base of the regolith.

B. PEDESTALS

(i) Description

Crickmay (1935) used the term pedestal rock to describe mushroom-shaped forms (see Chapter 9), but here *pedestals* are taken to be low, smooth-topped areas of granite that stand a few centimetres above the level of the adjacent rock surface. They are commonly surmounted by a boulder or block of granite. Some occur on slopes of gentle or moderate inclination (Fig. 8.14), but others stand above areas of sensibly flat platform. Examples have been noted at Domboshawa, in Zimbabwe, and in several parts of South Australia and Western Australia (Ucontitchie Hill, Mt Hall, Tcharkuldu Rock, Murphy's Haystacks in the former State, Kokerbin Hill and Hyden Rock in the latter).

(ii) Origin

Similar, though more dramatic, pinnacles due to the deep dissection of weak, unconsolidated or intensely weathered materials and described from various parts

Fig. 8.13 (b). The surface of the weathering front, exposed when the weathered rock was removed during the construction of a water conservation scheme at Kwaterski Rocks, was found to be dimpled due to the development of numerous shallow, saucer-shaped depressions.

of the world are commonly attributed to the protection of the preserved acicular columns by the slabs and blocks that cap them. Such an explanation does not adequately explain granite pedestals. Those located on slopes are bordered by arcuate shallow channels partly eroded by running water. At some sites (e.g. on Ucontitchie Hill, Fig. 8.14(b)), the flanking channels diverge upslope from the pedestal block. It cannot be argued that the divergence was caused by the obstruction, because there is no reason why the water flowing from above would not continue to the pedestal, and be diverted by it after scouring the up-slope side; of which there is no sign on any of the pedestals.

Discussing the particularly sharp nicks developed between platforms and backing flared slopes at Gorge Rock, near Corrigin, and on Hyden Rock, both in the southwest of Western Australia, Twidale (1968b) noted that small pools had been formed by the weathering effect of water that had dripped from the slopes above (Fig. 8.15). The pool depressions extended laterally, so that the base of the flare is slightly steepened.

The effect of such drip pools is even more dramatically, and in the present

230

Fig. 8.14 (a). Pedestal rock on Domboshawa, near Salisbury, Zimbabwe.

Fig. 8.14 (b). Pedestal rock on northeastern lower slope of Ucontitchie Hill, northwestern Eyre Peninsula.

Fig. 8.15. Small shallow depressions formed at the base of a flared slope (see Chapter 9) at Hyden Rock, southwest of Western Australia, by water trickling down the overhanging slope. Note the incipient linear depression formed parallel to the slope and level with the girl's feet.

context more relevantly, displayed at Tolmer Rock, in the Upper South East district of South Australia (Twidale and Bourne, 1976b). There, water dripping from a large residual boulder has, through excessive wetting and weathering of the platform on which it falls, caused the development of a series of shallow pools the pattern of which reflects exactly the plan shape of the boulder. These pools extend laterally and link to form a narrow, shallow moat around the base of the boulder (Fig. 8.16). Any run-off from upslope washes into the moat. Thus the moat is deepened and run-off from upslope is diverted. In this way, that area of the platform located beneath the boulder is not attacked by running water as is the rest of the exposed surface (Fig. 8.17). It is true that weathering proceeds beneath the perched boulder and the surface of the pedestal is also affected (Twidale and Bourne, 1977 - see also below), but this is slow compared with the effects of weathering and wash on the slope as a whole.

The same mechanism is adequate to explain pedestal development on platforms

232

Fig. 8.16. Moat surrounding large boulder at Tolmer Rock, Upper South East District, S.A.

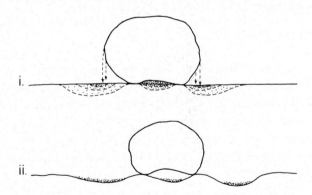

Fig. 8.17. Development of pedestal as a result of pool weathering around the base of a residual boulder.

but because of the soil and grus cover it may be more a matter of the weathering front penetrating deeper beneath the annular areas that receive moisture dripping from the boulder rim.

C. ROCK DOUGHNUTS

(i) Description

Rock doughnuts are raised rims surrounding rock basins (Fig. 8.18). They have been described from granitic environments in the Llano of central Texas (Blank, 1951a and b) and from central Eyre Peninsula (Twidale and Bourne, 1977). Annular rims are also developed in sandstone, for instance, in western Natal on the southern margins of the Drakensberg (Twidale, 1980a) and in coastal situations, as, for instance, near Shag Point, South Island, New Zealand (Twidale, 1976b, p. 377) and Cape Paterson, Victoria (Hills, 1971). If the absence of reference to these forms in the literature is any indication, however, well developed rock doughnuts are rare.

Those in central Texas are very large, being almost 3 m in diameter and standing 30 cm or so above the adjacent rock surface, and with a depression 80 cm deep within the rim. Those of Eyre Peninsula are much smaller. The coastal sandstone doughnuts are much larger than those observed in the Drakensberg of South Africa and the sandstone features differ from their granitic counterparts in that they are indurated with iron oxide or carbonate.

(ii) Origin

(a) Induration: Before concluding that he was unable to account for rock doughnuts, Blank (1951b) considered a number of possibilities. He wondered whether waters held in the pit had impregnated and indurated the surrounding

Fig. 8.18. Large rock doughnut located on the northern upper slope of Enchanted Rock, central Texas.

rock. He could find no supporting evidence from his study area in Texas, and the low porosity and permeability of the crystalline rocks in which doughnuts are developed renders the proposal inherently unlikely.

Moreover, iron itself is not highly soluble, but the reducing conditions that are produced by the many micro-organisms which inhabit pools of water that occupy rock basins are conducive to the dissolution of the material. After rain, the water could overlap on to the adjacent surface and, after desiccation, the iron salts could precipitate and then oxidise to give the observed red coloration. Thus, the rims bordering basins on Pildappa Rock and Ucontitchie Hill can, in some measure, be accounted for in terms of protection by a skin of iron oxide. Whether such marginal induration develops subaerially or under subsurface conditions is not known, but howsoever this may be, the best-formed doughnuts, at Enchanted Rock and Waddikee Rocks, lack such protection.

(b) Water scour: Blank speculated whether the flow of water over the surface of the granite domes he studied had been disturbed by the standing water in the basins in such a way as to protect the immediately adjacent rock surface, thus allowing the development of a raised rim. Experimental work and field observations concerned with both wind and water (see e.g. Allen, 1970, pp. 40 *et seq.*; Twidale, 1972a) confirm that such obstacles indeed interfere with flow and induce further turbulence, but the effect is to increase, rather than decrease, scouring and erosion. The granite and gneiss outcrops studied are, however, and to a greater or lesser degree, covered by foliose lichen and moss. Though lichens can cause rock disintegration (see e.g. Fry, 1927; Hale, 1967, pp. 162-163; Syers and Iskandar, 1973) the lichen present on the rims could also conceivably protect the surface, as required by Blank's working hypothesis.

If the areas marginal to rock basins were favourable to lichen growth because of their dampness, it could be argued that this ring of vegetation could divert flow and wash around the basin, induce increased turbulence and scour (see Allen, 1970, p. 40), and thus produce the annular rim that is essential to the formation of the doughnuts under investigation. Unfortunately for this suggestion, there is no evidence of pronounced lichen growth on the Texas doughnuts. Nor is the lichen and moss cover on the Waddikee doughnuts thicker or more complete than it is on much of the nearby rock surface. Moreover, there are no lichens and mosses on the rims at Pildappa and Ucontitchie, though this could be attributed to the iron patination, which could perhaps have inhibited their growth.

(c) Relief inversion: Blank (1951b) also considered whether the doughnuts could have originated as small, low topographic domes left behind in relief as remnants of circumdenudation by the development of gutters on the sloping granite surface. He visualised basins developing in the crests of these, leaving the resultant rims as rock doughnuts. The initiation of basins in such situations

and the development of doughnuts as a consequence of such local relief inversion
ought to be unlikely because water runs off upstanding areas. In special
circumstances, such as the fortuitous exposure of a concentration of weak
materials on the crest of an upstanding rock mass, such development can be
envisaged, but this must surely be a rare occurrence.

(iii) Evidence and argument

Despite these general arguments against such localised relief inversion
developing and giving rise to rock doughnuts, forms and processes which substantially corroborate this last of Blank's speculative suggestions have been noted
in the field.

First, in several places isolated blocks and boulders rest on plinths,
pedestals or small platforms of rock which, though in physical continuity with
the underlying granite, nevertheless stand higher than the adjacent slopes with
which they merge by way of rather steep inclines. Such pedestals have been noted
at Domboshawa, in Zimbabwe (Fig. 8.14 (a)) and on Ucontitchie Hill, Waddikee
Rocks, Wattle Grove Rocks, Mt Hall and Tcharkuldu Hill, all on Eyre Peninsula.
They can evidently form in the subsurface or under the influence of epigene
processes. At both Tcharkuldu Hill and Mt Hall angular blocks protruding from
the edges of thin layers of grus stand on bases which are a few millimetres
higher than the adjacent covered platform. This is presumably due to the blocks
protecting the rock beneath from the weathering effects of water down the joints.

On the other hand, there is no doubt that such platforms evolve under epigene
conditions. Thus as mentioned earlier in connection with pedestals, during
heavy rain run-off from the upper slopes of boulders at Tolmer Rock drips from
the outer edge to the platform below, splashes onto the platform, forming a ring
of pools, the pattern of which reflects the shape of the boulder. The pools
extend and merge to form a shallow, annular depression that becomes part of the
local drainage pattern. In this way a plinth develops beneath the boulder (see
above).

Second, boulders standing on pedestals commonly develop basal tafoni as a
result of moisture attack. Moisture also attacks the adjacent plinth or pedestal so that basins are initiated there. Allowing that the boulders must
eventually disintegrate, so exposing the basins set into the plinths, doughnuts
could be formed. If the hypothesis has any validity it should be possible to
find examples of weathering beneath boulders resting on plinths and of plinths
that display hollows.

At Tcharkuldu Hill and Waddikee Rocks, blocks of granite and gneiss respectively were moved from their natural positions on plinths revealing the surface
beneath. On that at Tcharkuldu Hill there was 7010 mm of moist debris, consisting of quartz particles, finely comminuted feldspar, a little clay and a

considerable amount of organic material. The granite beneath was moist and soft, being readily penetrated by the point of a geological pick, whereas the granite at the naturally exposed edges of the plinth remained hard. The underside of the block had shallow hollows, and, while no such depressions could be detected on the plinth, conditions were clearly ripe for their initiation.

At Waddikee Rocks, there was a similar debris layer (quartz, bleached mica, a little clay and organic material, in a shallow saucer-shaped depression, the floor of which stands some 4.3 cm below the level of the surrounding bedrock rim.

At several sites where boulders rest on plinths, hollows are obviously extending into the underside of the boulder and down into the plinth. In both there is less weathering at the edges of the blocks than toward the interior, giving rise to visors in the case of tafoni, and raised rims on the plinths. Thus, at Wattle Grove Rocks, a massive granite block with flared slopes rests upon a base which itself displays flared margins. The sheeting joint which separates the block from the plinth is convex upwards. Rock wedges, pitting indicative of subsurface weathering, and laterally elongate tafoni are developed along the fracture.

At Mt Hall several such plinth depressions occur beneath boulders. They extend in depth up to 8 cm below the bordering rim. Similar, though smaller, hollows have been observed between boulders and plinths at Ucontitchie Hill. At Murphy's Haystacks an hourglass-shaped boulder is separated from its plinth by a horizontal fracture (Fig. 8.19). A shallow tafone extends into the base of the boulder and a depression, some 10 cm deep and bordered by an irregular, but distinct, rim is developed in the plinth.

Finally, at Waddikee Rocks and Mt Hall there are plinths with saucer-shaped depressions, which can be regarded as incipient doughnuts.

Thus the deducible consequences of the working hypothesis involving relief inversion are found in the field. The explanation offered is consistent with the evidence that rock doughnuts are residual forms due to the weathering and erosion of a roughly circular zone inside them, and of an annular ring outside the rim.

Observations at several sites, and notably Tcharkuldu Hill and Mt Hall, suggest that pedestal development can be initiated at the weathering front, beneath the regolith. There is no reason why saucer development should not occur there. There is indeed evidence that basins are initiated in the subsurface (Twidale and Bourne, 1975a). Also the sandstone doughnuts of southern Africa occur on platforms that have manifestly only recently been exposed from beneath the regolith and there is no doubt as to their subsurface origin (Twidale, 1980a).

On the other hand, even if the process begins in the subsurface, it is certainly continued and developed under epigene conditions.

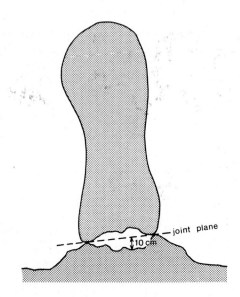

joint plane

10 cm

Fig. 8.19. Basal tafone and pedestal, Murphy's Haystacks, northwestern Eyre Peninsula, S.A.

D. RUNNELS OR GUTTERS

Both the flattish, upper surfaces and the steep sidewalls of bornhardts are frequently scored by channels cut in fresh granite. These forms are known by various names: *Rille, Granitrille, Silikatrille, Karren* (but see Bulow, 1942), *Pseudokarren, lapiés, lapiaz, cannelures, acanalduras* and so on. Here those on gentle gradients are referred to as *runnels* or *gutters,* from their similarity to the roof drains of houses. Those on steep slopes, on the other hand, are called grooves or flutings. Collectively the grooves and gutters are called channels. Joint-controlled forms (see Chapter 3) are known to some as *Kluftkarren,* slots or clefts.

(i) Description

Gutters are well developed on some river beds (Fig. 8.20). In addition, virtually all flattish granite surfaces are drained by systems of comparatively narrow, deep channels or gutters (Fig. 8.21). Most gutters are flat-floored and steep-sided, but some joint-controlled forms are V-shaped in cross-section. Some drain directly to the margins of the hill where the water runs over steep slopes either in linear channels or in thin sheets spread over the rock surface. Water draining to the margins of platforms disperses in the surrounding regolith. The channel sidewalls are commonly undercut, and some channels (as, for example, on Domboshawa, Zimbabwe, see Scott, 1967; and on Remarkable Rocks, Kangaroo Island) are bordered by raised, or what Lister (1973) called levee-like, rims. They are probably formed by water overflowing from the channels and depositing

Fig. 8.20. Gutters eroded in granite and exposed in the bed of the Ashburton River at Nanutarra, western Pilbara, W.A.

sufficient iron compounds to form a resistant veneer and to give rise to the raised rims. They are of non-structural origin and are comparable to the annular rim developed at the margins of some basins (see Twidale and Corbin, 1963; Twidale and Bourne, 1977, and earlier this chapter).

Such systems of connected channels develop only on extensive, gently sloping surfaces. Where relics of gutters are preserved on comparatively small blocks it is inferred that the host blocks are remnants of large radius domes or platforms that have disintegrated. Thus, near Bruce Rock of Western Australia, the channel remnants shown in Fig. 8.22 cannot have formed on such an isolated block. Similarly, several short and closely spaced gutters are preserved on the flattish top of a fretted or hoodoo rock (Fig. 8.23) standing in isolation on the crest of a low hill near Caloote, in the western Murray Basin in South Australia. The host block is about two metres maximum diameter, but is surely the remnant of a rock dome that has suffered extensive weathering and erosion.

Some gutters are blind and lead nowhere. Others link rock basins to form a rudimentary drainage system. Many of their courses are clearly guided by joints, though such structural influences are not as great as might be supposed. Some gutters, for example, run along joint traces for a few metres and then diverge

Fig. 8.21. Upper surface of Yarwondutta Rock, northwestern Eyre Peninsula, scored by shallow gutters cut in fresh granite and connecting rock basins (C. Wahrhaftig).

downslope. Others run across joints with no discernible diversion.

Gutters have been described from many climatic regimes, ranging from periglacial to tropical, and arid to humid.

(ii) Origins

That mechanical abrasion is effective in the erosion of the beds and sidewalls of channels is suggested first by the volume and velocity of water that, charged with mineral particles and other flotsam and jetsam, courses along the channels, second by the development of pot holes in some localities, third by the sinuous or meandrine form of some gutters which implies lateral erosion of the sidewalls, and last by the basal widening of some channels (though, see below, this is susceptible of alternative explanation), R. Dahl (1965) suggested that the channels excavated in granite in periglacial, recently deglaciated areas of northern Norway were eroded by subglacial streams flowing under high hydraulic pressure, with cavitation and corrasion both contributing significantly to the end result.

Fig. 8.22. Remnants of gutters preserved on isolated granite blocks near Bruce Rock, southwest of Western Australia.

On the other hand, the development of pitted surfaces (Twidale and Bourne, 1976a - see Chapter 3) in channel beds argues the effectiveness of chemical weathering there. Such microforms surely indicate a standing or retained moisture (depending on slope), the gradual loosening and detachment of the constituted rock particles, and their subsequent evacuation by the occasional flows of water. In addition, trickles of water rich in organic matter and emanating either from sheeting joints, or from the vegetated soil and weathered rock accumulated in the rock basins of upper slopes and surfaces also plays a part in the weathering and hence erosion of the channels. After desiccation, however, the organic material has at some sites had a protective influence sufficient to cause local relief inversion (see Chapter 9).

The flask-shaped cross-sections displayed by some channels may be due to longer wetting and hence more effective weathering of the bed.

Finally, and was also the case with rock basins, there is clear evidence that gutters are initiated at the weathering front. At several sites on Eyre

Fig. 8.23. Gutters preserved on fretted block near **Caloote**, western Murray Basin, S.A.

Peninsula gutters have been traced into the natural subsurface (Fig. 8.24). On both Dumonte and Crowder rocks, in the Minnipa-Wudinna district, what are separate channels on the exposed low dome converge and coalesce beneath the natural soils level (Fig. 5.23), just as do streams and channels on exposed surfaces; similar features have been observed in other lithological settings (Twidale, 1980a).

The reason for such subsurface extensions of channels is simple. Water pours into the regolith from the domes and platforms but is concentrated in the areas where runnels and gutters debouch from the hills on to the plains. Here the weathering front becomes lowered relative to the rest of the interface between altered and fresh rock. Shallow channels are formed aligned normal to the edge

Fig. 8.24. Gutter extending beneath the natural soil level (X-X) on the southern basal slope of Pildappa Rock, northwestern Eyre Peninsula.

of the outcrop. Where such piedmont zones have been exposed, pitting is markedly more pronounced in these linear zones than on the rest of the weathering front. Some such incipient channels fade to nothing only one or two metres away from the edge of the outcrop, but others extend several metres at least along the sloping contact between fresh and weathered rock.

After exposure the channels are enlarged and otherwise modified but some of them at least originate at the weathering front and are therefore intrinsically of etch character.

CHAPTER 9

FORMS ASSOCIATED WITH STEEP SLOPES

Inselbergs are characteristically steep-sided. In large measure this reflects the high dip of the orthogonal fractures that delineate the forms and of the sheeting joints that plunge steeply near the margins of compartments. In addition, the basal slopes of many bornhardts have been steepened and otherwise modified through the development of flared slopes, with which are associated tafoni (Chapter 10), platforms and scarp foot depressions.

A.   FLARED SLOPES

(i)   Distribution

Flared slopes (Fig. 9.1) have been noted or described from several parts of the world, but are apparently not developed in others, even at sites that appear to be suitable. In the Australian context they were first illustrated from Mt Buffalo, in the uplands of eastern Victoria (Hellstrom, 1941), and have been described from several parts of southern and central Australia (Twidale, 1962, 1964, 1968b, 1972b; Leigh, 1967; Dahlke, 1970) in areas ranging from the Kulgera Hills of central Australia to the southwest of Western Australia, the Mount Lofty Ranges and the South East of South Australia, the Geelong district of Victoria, and the Kosciusko and New England regions of N.S.W. Flared slopes are not well represented in tropical granite areas in northern Australia (the Devil's Marbles and Mount Bundey, near Darwin; north Queensland) and where they do occur they are poorly developed.

Flared slopes are present also in the Alabama Hills, in the eastern foothills of the Sierra Nevada near Lone Pine, California (Fig. 9.2); at high elevation (some 2500 m) within the Sierra Nevada near the Devil's Postpile; in the Cassia City of Rocks, Idaho (see Anderson, 1931); in the southern Appalachians (see Crickmay, 1935); on Agulhas Negras, in the Serra do Mantiquererra of south-eastern Brazil; from various parts of Zimbabwe (Fig. 9.3), including such domes as Domboshawa; in rudimentary form on Paarlberg and in the Kamiesberge of Namaqualand, both in Cape Province, R.S.A.; in the Sidobre of southern France (Fig. 9.4) at Palo Ubin, Singapore, and Tampin, West Malaysia, and so on. They occur in both interior and coastal sites (Figs. 9.5 and 9.6).

Flared slopes are not restricted to granite or granitic rocks. They are developed on sandstone and conglomerate, on limestone and on volcanic rocks: on the arkosic sandstone of Ayers Rock, on conglomerate at the Olgas, on quartzite

Fig. 9.1. Flared basal slope at Ucontitchie Hill, northwestern Eyre Peninsula, with a large flared residual boulder beyond.

in the southern Flinders Ranges, on limestone at Weka Pass, South Island, New Zealand, and in the southern Flinders Ranges, S.A.; on volcanic rocks (solvsbergite) at Hanging Rock (Mt Diogenes), Victoria; on basalt in Death Valley, California, and in northern Mexico; and on dacite in the Gawler Ranges, S.A. The common attribute of these seemingly varied lithological bases is that they are massive.

(ii) Description and Characteristics

Flared slopes consist of basal concavities up to 12 m high (Fig. 9.7). Some of the concavities are so pronounced that the slopes are overhanging. In places the concavity is not simple, but consists of two or three minor flares superimposed on the overall concavity (Fig. 9.8). The steepened slope is not everywhere disposed in horizontal zones, but rather follows the hill-plain or rock-soil junction (Fig. 9.9). Flared slopes attain their maximum development (that is, greatest degree of overhang and length of steep walls) in two topographic

Fig. 9.2. Flared boulders in the Alabama Hills (see Fig. 1.1) near Lone Pine, in the foothills of the eastern scarp of the Sierra Nevada, California.

situations. They are well developed on the points of spurs, as, for example, at Ucontitchie Hill (Fig. 9.10), and also in broad joint-controlled embayments, for example, Wave Rock, on the northern side of Hyden Rock, Western Australia (Fig. 9.7). They are found on both the northern and southern sides of residuals but are particularly well developed on the latter in southern Australia, though there are exceptions. They are developed on boulders as well as on larger residuals (e.g. Fig. 9.1; Twidale and Bourne, 1976b). They are preferentially exposed on the downslope side of hills and boulders as well as on larger the contemporary scarp foot zone, flared slopes also occur in subhorizontal zones well above the present plain level, as well as in joint clefts (Fig. 9.11). Excavations and bores have shown that they are incipiently present in the scarp

Fig. 9.3. Flared boulders at Bury Hills, Banket, Zimbabwe (W.D. Purves).

Fig. 9.4. These minor flares are developed on large residual boulders exposed in a quarry at Veyrières, near Castres, in the Sidobre of southern France. The flared sectors mark old soil levels affected by moisture attack.

foot zone beneath the present plains.

All-round attack of boulders or columns has produced conical residuals known as Chinaman's, or coolie, hats (Fig. 9.12). The acuminate boulders cited by Shaler (1887-8, p. 553) from Massachusetts are probably of similar origin, as are hourglass forms or dumb-bells, which vary from the pointed forms only in the depth of the concavity developed (Figs. 7.14 and 9.12 (a)). Hourglass forms have also been observed developed on basalt in Death Valley (Twidale, 1976b, p. 292). Blade-like forms and mushroom rocks are other variants and there are, in addition, several rocks that are named from their resemblance to specific objects, for example, anvil rocks (Fig. 9.12 (e)).

(iii) Origin

Several possible explanations can be suggested to account for flared slopes (see Twidale, 1962, 1971a, 1976b). The subaerial processes that can be cited as being possibly responsible for, or capable of shaping, the flares do not withstand close examination. Wave action, considered together with the likely age

Fig. 9.5. Flared slopes developed in arkose on the southeastern margin of Ayers Rock, central Australia.

Fig. 9.6. Flares developed along a prominent joint cleft on one of the Pearson Islands, Great Australian Bight (see Fig. 1.5).

Fig. 9.7. Wave Rock is 11-12 m high and is located in an embayment on the northern side of Hyden Rock, Hyden, in the southwest of Western Australia. The black streaks are due to algae and other organisms contained in water seeping from a storage located on the upper slopes of the Rock. Note the faint suggestions of double concavity developed on the lower slope.

of the forms, cannot account for their spatial distribution (for example, their occurrence at elevations of some 2500 m in the Sierra Nevada, 1600 m at the eastern margin of the Sierra, and 700 m at Ayers Rock and the Olgas of central Australia). Wind action does not offer a satisfactory explanation for either

Fig. 9.8. Multiple flares on southern slope of Pildappa Rock, northwestern Eyre Peninsula.

the local preferred orientation of the steepened slopes or, indeed, for their location in the scarp foot zone; the latter, being moist, is commonly better vegetated than the surrounding plain and is for that reason protected against any possible sand blast action. Running water fails to account for either the preferred orientation or their development on the points of spurs (where flow diverges). Crickmay (1935) attributed the mushroom rocks, some of which incorporate flared slopes, of the southern Appalachians to pronounced granular disintegration due to hydration on the low slopes shaded by the wider caprock.

Fig. 9.9. Flared slopes following inclined hill-plain, or rock-soil, junction at Chilpuddie Hill, northwestern Eyre Peninsula.

Quite apart from incorporating a circular argument, this explanation, like all the others mentioned, fails to account for the subsurface development of the forms. The only explanation that takes account of all the evidence is that the flares are a particular form of etch surface or weathering front developed in the scarp foot zone as a result of moisture attack on massive rocks and subsequently exposed (Fig. 9.13).

The crucial evidence is seen at such sites as Yarwondutta Rock and Chilpuddie Hill, both on northwestern Eyre Peninsula, and at Veyrières in southern France (Fig. 8.4), where incipient flared slopes, in the shape of concave sectors of the weathering front, are present beneath the natural land surface (Fig. 9.14). The material with which these bedrock forms are covered is not transported material covering flared slopes that were formed by exogene processes and then buried, but grus, or granite weathered *in situ*. The concavity results from the drying out of the surface and near surface soil and the persistence of moisture and hence longer duration of weathering depth.

This suggested mode of development explains why incipient forms can be found beneath the regolith, why flared slopes follow the rock-soil junction, and why

Fig. 9.10. (a) This prominent overhang is developed on a spur on Ucontitchie Hill. Note streaks of algal slime, and narrow basal platform. (b) At the Dinosaur, near Mt Wudinna, an earlier stage in the development of such undercut spurs is seen, for at its western end, the flares developed on the two sidewalls intersect.

Fig. 9.11. Flared slopes developed in a major joint cleft, Yarwondutta Rock, northwestern Eyre Peninsula.

they are preferentially developed on the shady or wetter sides of inselbergs and in relation to joints which permit the ready infiltration of water into the rock mass. Multiple flares are a consequence of repetitions of subsurface weathering and lowering of the plain (Fig. 9.13).

Flares are well developed on the points of spurs because there the fresh rock is attacked from two sides in the subsurface. The preferential occurrence of flares on the downslope side of boulders is due not so much to development as to exposure, for there grus is evacuated downslope, whereas on the upslope side detritus tends to accumulate. Initiation in the subsurface by moisture attack also explains why in northern Australia and other regions where flared slopes are rarely and poorly developed, the few examples occur on sites that are particularly moist. In the Georgetown area of north Queensland for instance boulders located near the channels of intermittently flowing streams tend to be the only ones that are flared. And so on - the field characteristics become comprehensible in terms of subsurface initiation.

Moreover, exceptions are understandable in terms of this working hypothesis. For example, at several sites such as the Dinosaur and Brazil (or Poondana) Rock, both on northwestern Eyre Peninsula, the major flares occur on the northern aspect of the residuals, but in every instance they are developed along major joints; the relative readiness with which water can infiltrate such zones more than compensates for the marked desiccation of the regolith on the exposed northern side of the hills.

In extreme cases the weathering penetrates laterally and the flare becomes a shelter or *tafone* (see Chapter 10), so that the latter merge with flared slopes as at Ayers Rock and at several sites on Eyre Peninsula and in Western Australia (Fig. 9.15).

Fig. 9.12. All-round attack of boulders by soil moisture has produced, above, (a) conical hills called Chinese or coolie hats, as seen here near Nonning on northern Eyre Peninsula and (b) a mushroom rock due to the pronounced basal sapping by soil moisture, Domeland area, Sierra Nevada, California (C. Wahrhaftig).

Fig. 9.12. (c) Above, mushroom rock due to sand blasting of weathered rock, southern Libya (R.F. Peel), and (d) below, acuminate blades on the slopes of Mt Manypeaks, southwest of Western Australia.

Fig. 9.12.   (e) Anvil rock, Caloote, S.A.

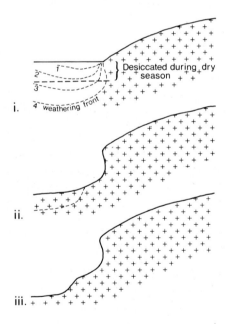

Fig. 9.13.   Stages in the development of flared slopes.

Fig. 9.14. The subsurface initiation of flared slopes is demonstrated by their presence in excavations, where they are seen to be a particular form of the weathering front; above, (a) seen in the Yarwondutta Reservoir where X marks the position of present and former soil levels, and (b) at Chilpuddie Hill, where the old soil line is clearly marked and where patches of grus remain. Both of these sites are near Minnipa, on northwestern Eyre Peninsula, S.A.

Fig. 9.15. In many places, as at Kokerbin Hill, in the southwest of Western Australia, flared slopes merge laterally with tafoni, suggesting that the latter are also initiated in the subsurface.

B. FRETTED BASAL SLOPES

Though lacking flared slopes as such, many boulders and some inselbergs display *basal fretting*: the lower sections of the steep bounding slopes of hills and the flanks of blocks and boulders are, to a greater or lesser degree, notched and overhanging (Fig. 9.16). The walls of the concavities are rough, not smooth and regular, as are flared slopes. This in many instances reflects the texture of the bedrock, the fretting being especially well developed in coarsely crystalline or porphyritic granite. Nevertheless, and despite morphological differences in detail, the fretting appears to be of the same origin as flared slopes, and is a manifestation of soil moisture attack in the scarp foot zone.

Many examples of basal fretting have been noted in tropical and subtropical arid and semiarid regions, where scarp foot water concentrations are highly significant. But they are by no means restricted to such areas. Thus, examples of locally flared or basally notched slopes have been illustrated from the gritstone tors of the English Pennines (Linton, 1964); the famous (or infamous)

Fig. 9.16. In coarse-grained rocks well-defined flares are not well developed. Instead, there is a zone of basal fretting as shown at the margin of Poldinna Rock, north of Minnipa, northwestern Eyre Peninsula.

Cumberland Stone, at Culloden, in northern Scotland, is so fretted (Fig. 9.17); there are granite boulders with rudimentary flares at Penninis Head, St Mary's, in the Scilly Islands. Also, several of the *menhirs*, or guidestones, of Dartmoor (Fig. 9.18) display both basal fretting and annular basal depressions, some with pools of water. The basal fretting around the menhirs argues a rapid rate of development, at least in the geological context.

In some few cases, basal indentation occurs at the margins of ephemeral lakes and may be caused by weak wave attack and/or weathering by saline waters. Jutson (1917, 1934, p. 241) described what he called billiard-table surfaces backed by basally notched cliffs from the Salinaland division of Western Australia (Fig. 9.19) and similar forms have been observed in dacite and schist respectively at the margins of Lake Gairdner and Lake Greenly, both in southern South Australia.

But these are exceptions. Most basal fretting is caused by soil moisture, and the resultant notches, like flared slopes, provide a measure of recent soil erosion.

C. ROCK PLATFORMS

(i) Description and distribution

Flared slopes extend laterally into gently sloping bedrock surfaces called *rock platforms* (Twidale, 1978a; see also Chapter 7 C (i) (b)). They vary in width between a few centimetres at the base of some boulders, and a few metres (Fig. 7.11) and several hundreds of metres, as, for instance, at Peella Rock

Fig. 9.17. The Cumberland Stone, from which the Duke of that name directed operations during the battle of Culloden Moor in 1745, displays distinct basal fretting (Brit. Tour. Auth.).

and Corrobinnie Hill (Figs. 1.26 and 7.10) in South Australia, and the Humps and Varley Township Hill in Western Australia (Fig. 9.20). They vary in area from a few square metres to several square kilometres. They are most characteristically developed bordering residual hills, but they are also found in isolation, unrelated to any upland mass, though whether this has resulted from the elimination of a former inselberg or whether there never was an upland and the platforms are merely crestal exposures of very large radius concealed or incipient domes, is a moot point.

In detail, the platforms are dimpled, due to the development of shallow, saucer-shaped depressions, and scored by gutters that, together, form a rudimentary drainage network (see Fig. 8.21). Blocks, boulders and patches of grus remain on some platforms.

Small platforms in gneiss or granite have been noted adjacent to inselbergs at Shashe, Gokomere and at other sites in Zimbabwe, as well as in central Namibia and in the Mojave Desert of southern California. Extensive granite platforms are developed adjacent to the Hoggar Mountains of southern Algeria (R.L. Folk, pers. comm.). Similar forms are developed in arkose adjacent to Ayers Rock, in

260

central Australia, though they are here of only limited extent (Twidale, 1978b).

(ii) Origin

Platforms are erosional in the sense that they cut across jointing and other rock structures. Like flared bedrock slopes, they have been shown to extend beneath the present regolith and, also like flares, they are regarded as etch forms, or exposed parts of the weathering front. Indeed they are merely lateral extensions of flared slopes. They are especially well developed in wet sites such as the scarp foot zone, topographic lows (e.g. the Corrobinnie Depression - see Bourne *et al.*, 1974) and bordering ancient watercourses or lakes (as in the southwest of Western Australia). The grus and boulders found on some platforms are remnants of the regolith that formerly covered the planate forms and that has now been partially removed.

Fig. 9.18. This menhir, or stone guide post, on Dartmoor has a depression around its base due to weathering and wash by water dripping from the stone. The post itself is fretted at the base, again because of water attack.

D. SCARP FOOT DEPRESSIONS

(i) Description

Shallow moats or linear depressions (Cotton, 1942, p. 94), also known as

Fig. 9.19. This undercut basal slope of a large residual boulder at Balladonia, on the edge of the Nullarbor Plain in the southeast of W.A., is due to lake waters weathering the rock and slight wave action removing the loosened fragments.

Fig. 9.20. Platform bordering the Humps, southwest of W.A.

Bergfussniederungen and *depressions de piedmont*, are found around the bases of some inselbergs in arid and semiarid (especially savanna) landscapes (Fig. 9.21). Such *scarp foot depressions* have been reported from West Africa (Thorbecke, 1927; Passarge, 1923, 1928; Sapper, 1935, pp. 105-107; Pugh, 1956; Clayton, 1956), and from the Sudan, where they are known as *fules* (Ruxton and Berry, 1961), from the Egyptian Desert (Dumanowski, 1960). They occur in central and southern Australia (Mabbutt, 1967; Twidale *et al.*, 1974), and in the Mojave Desert of southern California (T. Oberlander, pers. comm. - see Fig. 9.22). Like other piedmont forms, they are developed in settings other than granitic (e.g. Twidale and Bourne, 1975c).

Most of the moats are just a few metres across, but some attain widths of several scores of metres, and that around Gebel Harhagit, in Egypt (Fig. 9.23), is more than a kilometre wide in places (Dumanowski, 1960). Most are only a metre or two deep, but in others the floor stands several metres below the level of the outer rim.

(ii) Origin

Dumanowski (1960) favoured a structural (lithological) origin for scarp foot depressions. Certainly the specific features he described from the Egyptian Desert are eroded in metamorphic rocks, whereas the backing ridge consists of granite. In many other instances described in the literature, however, no such lithological contrast can be detected. On the contrary, the bedrock in the floor of the depressions is apparently of the same type as that exposed on the adjacent hills.

Minor scarp foot depressions are undoubtedly caused by runoff scouring the inner fringe of the regolith where it laps up against the bare rocky hillslope, as, for instance, at Chilpuddie Rock, on northwestern Eyre Peninsula. Cotton (1942, pp. 94-95), utilising the field accounts of such workers as Thorbecke (1915) and Holmes (1918), pointed out that rivers in some arid and semiarid regions 'consistently hug the mountain bases in a manner than cannot be fortuitous' (Cotton, 1942, p. 42) and goes on to suggest that the linear depressions are river channels. There is here a possible confusion of cause and effect. Moreover, accepting the notion of scarp retreat, Cotton proceeded to argue that the mountain fronts and river channel depressions must recede together. This hypothesis is difficult to sustain, as both Pugh (1956) and Clayton (1956) pointed out, though the former continued to assert that the margins of granite domes are subject to scarp retreat. Very few of the depressions observed in Australia and north Africa, are occupied by streams. In those that are, it may be a case of the drainage channels running in the depressions rather than excavating them. Furthermore, some of the depressions stand at the base of residuals too small to generate streams, and others occur where no streams debouch from the uplands.

Fig. 9.21. (a) Minor depression, only some 15-16 cm deep, at the base of a flared
slope on Yarwondutta Rock, northwestern Eyre Peninsula. On the backing slope
the flutings are cut in fresh granite but the divides between are not only
coated with desiccated organic material but lead upslope to gutters cut in the
gently sloping upper surface of the residual. These present ribs were formerly
the floors of flutings which, protected by the organic material, resisted
erosion, so that the divides were worn away to become the present grooves: an
example of local relief inversion.

The most plausible explanation for scarp foot depressions is that due to

Clayton (1956) and supported by Twidale (1962, 1967) and Bocquier et al. (1977).

Runoff from the hills saturates the scarp foot zone and, in consequence, intense

Fig. 9.21. (b) Scarp foot depression at Wattle Grove Rocks, near Wudinna, northeastern Eyre Peninsula.

Fig. 9.21. (c) Annular depression developed around a granite nubbin, central Australia (Div. Land Res., C.S.I.R.O.).

Fig. 9.22. Sketch of scarp foot depression in the Mojave Desert, southern California (drawn from photograph supplied by T. Oberlander).

deep weathering occurs there. There is ample evidence of such chemical attack in a wide variety of lithological settings. The fines produced by the weathering of the country rock may be flushed out (Ruxton, 1958), or there may be a volume decrease, compaction and settling, and surface subsidence (Twidale, 1962). Additionally, and as doubtless happens in some places, once formed the depressions may have been deepened by intermittent or episodic surface streams. Whatever agent is responsible, however, the scarp foot zone is lowered and a topographic depression aligned along the hill base is formed.

Bocquier *et al.* (1977) regard the forms as having developed in the humid tropics. They believe that where these scarp foot depressions are found in arid lands they are inherited from former humid climatic phases, but this suggestion is sustained neither by general argument nor by the field evidence. For example, water is, if anything, relatively more important in arid and semiarid lands than elsewhere and, because of its concentration in the piedmont, achieves results that are more pronounced than in other climatic contexts.

E. THE PIEDMONT ANGLE

One further overall result of the scarp foot weathering and erosion manifested in the several landforms just described is that the *piedmont angle*, or nick, (Rahn, 1966; Twidale, 1967) becomes even more pronounced (Figs. 7.6 and 9.24). The feature is basically a structural form (Twidale, 1967) and in granitic terrains is roughly coincident with the margins of massive compartments. But

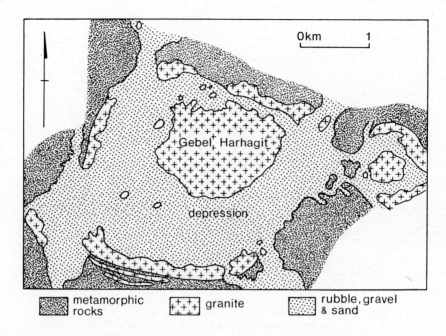

metamorphic rocks   granite   rubble, gravel & sand

Fig. 9.23.   Scarp foot depression at Gebel Harhagit, eastern Egypt (after Dumanowski, 1960).

Fig. 9.24.   The piedmont angle developed at the base of a nubbin near Naraku, northwest Queensland.

weathering and surface lowering due to volume decrease, as suggested by Ruxton (1958), or preferential erosion of the zone of altered rock, cause the structural feature to become more pronounced. Truly angular nicks are particularly commonplace in granitic rocks where vertical or near-vertical fractures are exploited and exposed by weathering and erosion.

F. GROOVES OR FLUTINGS

Many steep slopes, including flared slopes, are scored by *grooves*, and are said to be *fluted* (Fig. 1.23).

(i) Description

Most grooves are U-shaped (Fig. 9.25) in cross-section (cf *Rundkarren* formed in limestones), others are more open, while many others are broad, shallow and flat-floored.

Flutings are well developed in equatorial areas of consistently high rainfall (Fig. 9.26). They were described from Singapore and Malaysia (Logan, 1849, 1851) in the middle of the last century, and Logan (1848, p. 102) recounted how:

Fig. 9.25. U-shaped flutings on Frog Island, near Palo Ubin, Singapore.

Fig. 9.26. Fluted boulder near Tampin, West Malaysia.

   Amidst the luxuriant forest that always covers granitic hills and mountains,
the explorer suddenly finds himself facing a high perpendicular wall of rock,
indented by numerous vertical grooves sometimes 5 or 6 feet in depth.

The humid tropical provenance of some of the forms has seemingly been confirmed
by several workers (e.g. Helbig, 1940; Hitchcock, 1947; Alexander, 1959;
Birot, 1958; Bakker, 1958; Tschang, 1961, 1962; Wall and Wilford, 1966).
   On the other hand, these channels are found far beyond the confines of the
humid tropics. They have been described from such arid and semiarid areas as

central Brazil (Branner, 1913), and occur also in the arid lands of southwestern Angola (e.g. Jessen, 1936), on Domboshawa in Zimbabwe (Scott, 1967), on Paarl Mountain near Capetown (Fig. 3.28) and the Volzberg in the savannas of Surinam (Bakker, 1958). They are well developed in several sites in Victoria and Eyre Peninsula (Figs. 1.23, 2.29 and 9.27). They are found in central Australia, as, for instance, on Ayers Rock (Twidale, 1979b), but are strangely lacking in northern Australia; they are not wholly absent, but they are rare, and then poorly developed. They occur on cool, high altitude areas in the highlands of tropical Brazil (Fig. 6.4). They have been described from recently deglaciated or cold areas such as Scandinavia (R. Dahl, 1965) and Bohemia (Demek, 1964a and b), from seasonally cold areas like the Margeride of central France (Godard *et al.*, 1972).

Apart from congeners in limestone, flutings are also quite commonly developed on sandstone, in the Roraima Plateau of Venezuela, in Germany, southern Africa, central Australia and southern Brazil (e.g. Klaer, 1957; White *et al.*, 1966; Urbani, 1977; Twidale, 1978b, 1980a; J. Bigarella, pers. comm.).

Fig. 9.27. Fluted granite slopes on a large block at Granite Rock, near Geelong, Victoria.

(ii) Origins

(a) Structural factors: That a broad and general control is exerted by structure on the development of flutings was first indicated by Branner (1913) who
pointed out that the forms are developed only on massive rocks. They do not
occur on fractured bedrock, save in special circumstances, presumably because
the moisture that is most likely responsible for their development can there
percolate into the cracks and crevices and does not attain the volume or velocity required to scour isotropic rock surfaces.

On the other hand, some grooves are *Kluftkarren* (Carlé, 1941; Schmidt-Thomé,
1943), being due to the exploitation of fractures by weathering agencies. Logan
(1849, 1851) was fully aware of the possibility of structural control in detail
and suggested that some of the grooves he observed on Palo Ubin in the Johore
Strait may have been developed in linear zones where the granite was less cohesive than elsewhere. He was unable to find field evidence to support this
suggestion, but In other places, as, for instance, at Granite Rock near Geelong,
Victoria, and in Andorra, the coincidence of grooves and joints is quite obvious.
The near-vertical bedding of the Ayers Rock arkose is also clearly exploited.

But the control exerted by structure is not absolute. Gutters and grooves on
Mt Wudinna and Ucontitchie Hill, northwestern Eyre Peninsula, follow along joints
for short distances but run in fresh, massive rock for the most part, even
though joints are present nearby. One gutter on the lower slopes of Mt Wudinna
runs diagonally across a major joint.

Also on Mt Wudinna, some grooves have been eroded preferentially in the
porphyritic granite that comprises the greater part of the residual, rather than
in locally developed aplitic veins and lenses, though they have preferentially
exploited the junction between the two rock types, and where this has occurred,
the gutters are bordered by raised rims of aplite.

Thus, although there is a general structural control of distribution of
grooves in some areas, most are developed independently of structure in detail.

(b) Processes at work: Most workers have concluded that grooves and gutters
are due partly to chemical and partly to mechanical processes, and debate on the
origin of the channels has revolved partly around the relative significance of
mechanical abrasion and chemical weathering, particularly by moisture, and
partly around the role of biota, especially such plants as lichens.

Branner (1896, 1913) noted and investigated flutings, some of them 2 m deep,
developed on granite and syenite residuals in Brazil. He attributed them
largely to 'a narrow localisation of mechanical action by water' and attached
greater importance to this process in the later than in the earlier paper.
Others have followed Branner in emphasising mechanical work. Klaer (1956), for
example, considered both mechanical and chemical processes but, because both

bed and sidewalls of the channels are smooth, emphasised abrasion by running water. Tschang (1961) believed that the *Pseudokarren* of Palo Ubin belie their name and are due to wash, and both Scott (1967) and Dragovich (1968) made similar interpretations of the gutters of Domboshawa, near Salisbury, Zimbabwe, and Remarkable Rocks, Kangaroo Island, South Australia, respectively.

Others, such as Ule (1925), attribute grooves wholly to chemical weathering. Scholtz (1946), p. xlix) noted runnels on granitic inselbergs at Vredenburg and Witteklip, in the western Cape Province of South Africa, and thought them due to 'solvent action of downward trickling solutions charged with cyclic salts and organic acids ....'. The latter, he thought, derived from soil-filled rock basins developed in the upper slopes of the residuals. Support for such an origin involving the dominance of the chemical action of water comes from the development of runnels on the faces of blocks and boulders that generate insignificant volumes of runoff; from their occurrence the interior walls of hollows and on overhanging faces draining very small areas (Fig. 9.28); and from the pitting (Chapter 3) found in the beds of several of the channels (Fig. 9.29).

But most workers have steered a middle course and attributed runnels to both mechanical and chemical agencies.

The weight of opinion, however, favours chemical processes and especially solution. Flows of water cause the rock to be wetted in linear zones extending downslope; chemical reactions take place; subsequent flows of water remove the particles loosened by weathering and so a linear depression or channel develops (see, for instance, Birot, 1958, pp. 24-25; Demek, 1964a and b). Once formed, the channel tends to gather water and hence the wetting and erosional processes are augmented.

As for biotic effects, several agents and processes have been noted. Alexander (1959), for instance, considered lichens to be all-important. She argued that lichens colonise moist linear zones and that, by retaining moisture, they contribute materially to channel development. Lichens are capable of both mechanical and chemical weathering (Fry, 1924, 1926, 1927; Syers and Iskandar, 1973), but it is surely more likely that they colonise zones that are already moist rather than growing in linear patterns in the first place. For this reason, most workers assign to lichens, mosses and other organic agencies a contributory rather than dominating or initiating role. Moreover, in some circumstances organic (algal) slime has a protective function. Rills rich in organic matter initially increase the rate of weathering and give rise to intensely pitted channel floors. On the other hand, this same organic slime, on drying, has a protective effect on the rock surface and has produced localised relief inversion at several sites like Yarwondutta Rock and Pildappa Rock, near Minnipa, and the Dinosaur, near Wudinna (Fig. 9.21 (a)).

Fig. 9.28. (a) Shallow flutings developed on the overhanging face of a granite block at Remarkable Rocks, Kangaroo Island, S.A.

Fig. 9.28. (b) Shallow flutings on boulder on Ucontitchie Hill.

Fig. 9.28. (c) Shallow flutings in a tafoni developed on a flared granite boulder at Murphy's Haystacks, western Eyre Peninsula.

Some channels are merely continuations of the gutters draining upper slopes, but some are confined to the steep slopes. Pot holes indicate that scouring and abrasion are active on these steep gradients, but the very steepness and over-hanging character of some of the bounding walls of the granite residuals seem-ingly precludes simple abrasion as a significant factor because the flows become separated from the bedrock surface.

In these circumstances a number of other factors may come into play. Separa-tion of flow may result in or lead to collapse and impact in linear zones - the so-called 'water-curtain' effect. In conditions of high velocity flow, cavita-tion may occur, with high pressure waves and high speed water jets causing local rock shattering: this may be how the pot holes and other depressions found

Fig. 9.29. Channel with pitted floor stained with dried algal slime, eastern side of Ucontitchie Hill, northwestern Eyre Peninsula, S.A.

along some grooves, as, for example, on Yarwondutta Rocks (Rig. 9.21 (a)), are initiated. Scouring could cause increased surface roughness and induce further perturbation, air entrainment and turbulence, and hence increased erosion, all in linear zones running down the steep bedrock surfaces.

But, as with gutters, the field evidence suggests that, important as free flow is in the evolution of these grooves, trickles of water are also significant. Such trickles, some deriving from patches of soil, other from sheeting joints, persist long after rain has ceased, even on overhanging slopes to which, presumably, they adhere by surface tension.

(c) Coastal developments: Flutings have also been described from a number of modern coastal sites at Palo Ubin (Figs. 4.15 and 9.25), in the Seychelles (Bauer, 1898), Banka (Helbig, 1940), northwestern Spain (Carlé, 1941) and on Kangaroo Island (Dragovich, 1968). In some of these places they extend into the intertidal zone. On southern Yorke Peninsula, grooves have been exhumed

from beneath a cover of late Pleistocene aeolianite (wind-blown calcarenite or
dune sand).

The evolution of these coastal channels appears to be similar to that adduced
for inland sites: spray as well as rainwater runs down the surface of massive
blocks and, by a combination of chemical and mechanical attack, channels are
developed. The waters are, of course, highly alkaline, which, according to
some writers (Joly, 1901; Mason, 1966, p. 142; Bakker, 1958; Twidale, 1980a),
facilitates the weathering of acid rocks like granite. The development of grooves
on overhanging and sheltered slopes at Remarkable Rocks, Kangaroo Island (Fig.
9.28 (a)) suggests that, as at inland sites, trickles of water and the chemical
weathering they achieve are largely responsible for the development of some
grooves. At the same site, some gutters are flask-shaped in cross-section and
there is sand in the floors of the channels, indicating that these are most
likely formed by mechanical abrasion, so that in coastal sites, as elsewhere,
various processes are at work forming and shaping these narrow linear channels.

(iii) Subsurface initiation

The arguments concerning the relative importance of mechanical and chemical
processes in the development of gutters is interesting *per se*, but are also
linked with the question of their initiation, for there are strong indications
that some of the forms, at least, originated at the weathering front.

In 1846 J.R. Logan observed and investigated deep grooves developed in granite
on Palo Ubin, an island at the eastern end of the Johore Strait (Logan, 1849,
1851). He reported similar features from uplands in the southern part of the
Malay Peninsula (Logan, 1848, p. 102). On Palo Ubin many of the granite blocks
exposed on the beach were fluted and grooved, and inland many vertical and near-
vertical faces of the blocks were scored and grooved, some of them flask-shaped
in cross-section, and some deep and narrow, being about 0.76 m deep and only
0.61 m across (Logan, 1848, 1849).

For various reasons, Logan considered contemporary processes of weathering
and erosion inadequate to explain the grooves, but he made one crucial observation and several comments germane to the present discussion. He noted that some
of the grooves of Palo Ubin extend beneath the soil cover. Unless there had
been exposure, development of flutings, burial and subsequent exhumation, a
sequence of events unsupported by any evidence Logan could observe, this would
appear to demonstrate that the flutings had been initiated below the soil cover.
Nevertheless, Logan remain uncommitted in his 1849 paper. By 1951, however, he
asserted categorically not only that the grooves but all the boulders and blocks
he had observed on Palo Ubin together constituted what would today be called an
etch surface or complex weathering front. Thus Logan wrote:

> If ... we conceive the external layer of the island, when it first became
> exposed to decomposition, to have resembled in character the zone that has

been laid open for our inspection ... it is easy to comprehend how the
wasting away of the more decomposable parts might at last leave exposed
masses, including bands of the less stubborn material already partially
softened or disintegrated under ground, and that the action of the atmosphere and rain-torrents would gradually excavate the more yielding portions
until the solid remnants exhibited their present shapes. (Logan, 1851,
p. 328)

Thus Logan was clearly advocating a subsurface origin, development at the
lower limit of weathering or weathering front. In general, however, the opposite
view has prevailed. Working at the same general area and same site, respectively, Alexander (1959) and Tschang (1962) concluded that the gutters formed on
exposed rock surfaces. The generally held view was expressed by Wilhelmy (1958,
p. 199) when he stated that flutings in the humid tropics evolved *auf freilie-
genden Gesteinsoberflächen*.

It has taken more than a century for Logan's suggestion of a subsurface
origin to be revived. Bremer (1965) suggested as much with respect to the
arkosic Ayers Rock, and evidence from Eyre Peninsula (Twidale, 1971a, p. 90;
1976b, pp. 203-204; Twidale and Bourne, 1975a) clearly demonstrates that several
minor granite landforms, including the gutters and basins of gently sloping rock
surface, have been initiated at the weathering front (Figs. 5.23, 8.11, 8.12 and
8.21), and strongly suggests that the process continues (see also Boyé and
Fritsch, 1973).

Next, although there is so far no direct evidence, apart from that cited by
Logan, that flutings are initiated in the subsurface - no excavations in which
steep grooved walls have been exposed have been reported - there is much indirect
evidence. The grooves developed on boulders as well as those developed on the
flanks of larger residuals are commonly associated with flared slopes, which are
demonstrably exposed weathering fronts. Such associations are clearly displayed
at Granite Rock, in Victoria, at Ucontitchie, at Scholtz Rock, and Podinna Rock,
all on Eyre Peninsula, and at Kokerbin Rock, in Western Australia.

Second, in several areas it has been noted that channels have a restricted
vertical distribution or zonation. At Vredenburg, near Saldanha, in the western
Cape, South Africa, fluting is restricted to two horizontal zones, the first of
which is located just below the flattish upper surface of the dome, and the other
below a distinct shoulder about halfway up the slope of the residual (see Pl.
III, Fig. 1, in Scholtz, 1947 - also Fig. 9.30). Both zones lie just below what
can be interpreted as old erosional surfaces and in both instances soil moisture
attack is implied. On Mt Wudinna, on Eyre Peninsula, the flutings are in some,
though not all, instances continuations of gutters draining the flattish upper
slopes, but they are, in any case, confined to a particular flared steep slope
located on midslope and associated with a former piedmont zone (Twidale and
Bourne, 1975b). At both Granite Rock, in southern Victoria, and in the Tampin

Fig. 9.30. Smooth upper convex slope with finely fluted lower slope, coincident with flares and steepening, on inselberg area Vredenburg, western Cape Province, R.S.A. Another set of vertical pipes with wider chambers is preserved on a high cliff beyond, suggesting another, earlier, phase of development of the inselberg and of flutings.

region of West Malaysia (Fig. 9.31) grooves occur on upper slopes of boulders that are roughened or pitted (Twidale and Bourne, 1976a) as a result of subsurface moisture attack. All of this suggests that the flutings have been initiated in the subsurface.

Thus, many grooves and gutters are, as Logan (1849, p. 6) put it, 'prolonged beneath the ground'. Similar *Karren* of subsurface origin are also developed on limestone. There is no suggestion of burial and exhumation following development. On the contrary, it is grus, or weathered granite, which has been excavated to reveal the scored weathering front. These forms are therefore initiated beneath the soil surface and have been merely enlarged and modified after exposure.

Fluted surfaces are but one of several granite forms that appear to have been initiated at the weathering front. If correct, this has considerable implications for the question of zonation of granite forms in particular and landforms in general. It can be argued that grooves are especially well developed in humid tropical lands (though the forms are not everywhere present in such regions) and that those examples reported from arid and cold

Fig. 9.31. Boulder near Tampin, West Malaysia, with flutings restricted to upper slopes.

lands are relics from former ages when different climatic conditions prevailed. This suggestion calls for complex climatic changes for which there is at present little or no evidence. Another and more appealing possibility is that humid warm conditions obtained widely through much of the Mesozoic and earlier Cainozoic, that is during the period when runnels and grooves appear to have been forming in the subsurface. Thus in southern Australia, both the age of the forms and what is known of the climatic history of the areas concerned suggest development under humid tropical conditions. Palaeontological evidence suggests that humid warm conditions obtained in central and southern Australia during the earlier and middle Tertiary (see Dorman, 1966; Ludbrook, 1969; Callen and Tedford, 1976; Twidale and Harris, 1977; Kemp, 1978). Consideration of the denudation history suggests that many of the grooves found on the flanks of such granite residuals as Yarwondutta Rocks, Pildappa Rock, the Dinosaur, and so on, as well as many of those on the lower slopes of Ucontitchie Hill, are of early

Cainozoic age (see Twidale and Bourne, 1975b); those on the upper slopes of Ucontitchie and of Mt Wudinna are evidently older. All may have been initiated in the subsurface by streams of water pouring from the bare rock slopes into the soil and grus, and running down the concealed but real concave (flared) weathering front.

In this context the fluted surfaces of northern Scandinavia and Bohemia may be such forms initiated beneath and to some degree preserved by, a regolith,but exposed by glaciation. Here again, however, there are curious anomalies for as far as is known no gutters or grooves have survived, if they ever existed, on Dartmoor and other granitic uplands of Britain, even though warm humid conditions obtained through the earlier Tertiary on Dartmoor for instance (Reid, 1913).

Again the fluted surfaces of the deserts may have formed by groundwaters at the weathering front either under present conditions or in earlier savanna or humid tropical environments but again as a consequence of moisture attack at the weathering front. Soil moisture and groundwaters are so widespread as to be virtually ubiquitous, so that climatic conditions above the ground became largely irrelevant in the context of landform initiation. Those forms originating at the weathering front are azonal in terms of present conventionally defined climatic regions.

CHAPTER 10

CAVES AND TAFONI

Caves are underground openings connected with the atmosphere. Granite masses
are not noted for cave developments, but several have been reported from various
parts of the world. *Tafoni*, or relatively shallow hollows, on the other hand,
are widely developed on granitic rocks.

A. CAVES ASSOCIATED WITH CORESTONES AND GRUS

Occasionally the preferential subsurface flushing or evacuation of friable,
weathered rock has left irregular, tubular voids, largely defined by corestones
set in grus. Such subsurface erosion is achieved by streams that are diverted
underground and re-emerge a short distance downstream.

Thus Labertouche Cave, near Neerim South, in Victoria, is about 200 m long
and essentially straight, though irregular in detail. It begins in a sinkhole
in a blind valley and its point of emergence is marked by a pronounced re-
entrant (Ollier, 1965), and others of the like kind have been reported from the
uplands of eastern Victoria and adjacent parts of N.S.W. (Finlayson, 1981).
Others occur in central and southeastern Queensland (Shannon, 1975; B. Finlayson, pers. comm.) and in parts of the Western Cordillera of the U.S.A. They
present the same hazards as true karst caves: in 1980 a caver died from injur-
ies received in a fall and immersion in a 'boulder cave', one of the Lost Creek
Granite Caves, in Colorado, caves which, though up to 15 m high, are also
characterised by boulder falls and by streams that from time to time run at high
velocity (Arnold, 1980).

Feininger (1960) reported what he called pseudokarst forms including caves
developed in acid plutonic rocks in Columbia, and another cave system is developed on the granitic Makatau inselberg in the Rupununi savannas of Guyana
(Shaw, 1980), where the form of the openings in plan is clearly influenced by
the massive orthogonal joint system. They run for some 60-70 m between corestones. According to Shaw they have been exhumed (surely revealed?) during the
later Cainozoic and to have developed during an earlier arid phase.

These caves appear to be due to the piping of flushing of grus. The corestones and grus, either remain sufficiently rigid to form a roof and sidewalls,
or the corestones subside toward the evacuated area, but are packed closely
enough to jam together and form a roof or reasonable stability.

Again, some ten granite caves, varying in length between two and fifteen metres, have been reported from the High Tatra Mountains of Poland (Wojcik, 1961a). Some appear to be of the same origin as those already considered, but others are attributed to preferential solution acting along sideritic veins. There are also vertical shafts and niches in the Karkonosze Mountains, where they are associated with feldspar-rich zones. Calcite encrustations are developed on the ceilings of some of these caves (Wojcik, 1961b).

B. CAVES ASSOCIATED WITH FRACTURES

Weathering along sheeting planes has in many places created gaping partings. At a few sites such weathering has proceeded to such an extent that the openings are sufficiently large to be called caves. Thus the Enchanted Rock caves, located on the bornhardt of the same name in the Llano of central Texas, are developed at two levels (Kastning, 1976). The upper cave is some 250 m long (Fig. 10.1), the lower about 70 m. Each is 2-3 m high. Both follow along major sheeting joints and appear to be due to the weathering of the granite by moisture, with the greater part of the cave development taking place on the underside of the upper sheets (Fig. 10.2).

Boone's Cave, in North Carolina, is similar. Developed in a granite-gneiss, it is some 40 m long, displays obvious joint control in plan, and is thought to be essentially 'an enlarged bedding plane' (Hedges, 1978). Similar fracture-controlled caves in granitic rocks, and totalling some hundreds of metres in length, are developed in California.

Caldcleugh (1829) reported silica stalactites in gneiss "caves" near Rio, but the latter appear to be merely deep overhangs due to differential movement of gneissic masses and are not necessarily due to weathering and erosion.

C. TAFONI

(i) Description

Tafone (plural *tafoni*) is an Italian word meaning, variously, a perforation or a window, depending on regional usage, but as applied in geomorphology it means a shallow cavern or a hollow (Figs. 1.24 and 10.3) partially enclosed through the preservation of a visor. It was first used in a geomorphological context by Penck (1894, p. 219) of forms he had observed in Corsica. They are especially well developed in granite rocks though they have formed also in sandstone (e.g. Ollier and Tuddenham, 1962; Michel Mainguet, 1972; Twidale, 1978b; Johnson, 1974) and various metamorphic rocks (e.g. Martini, 1978). In the granite context tafoni are particularly common in the scarp foot zone, but are widely developed on granite outcrops generally. Shelters lacking visors and due to basal sapping, to the exploitation of inter-secting fracture sets (Fig. 10.4), or to preferential weathering are also

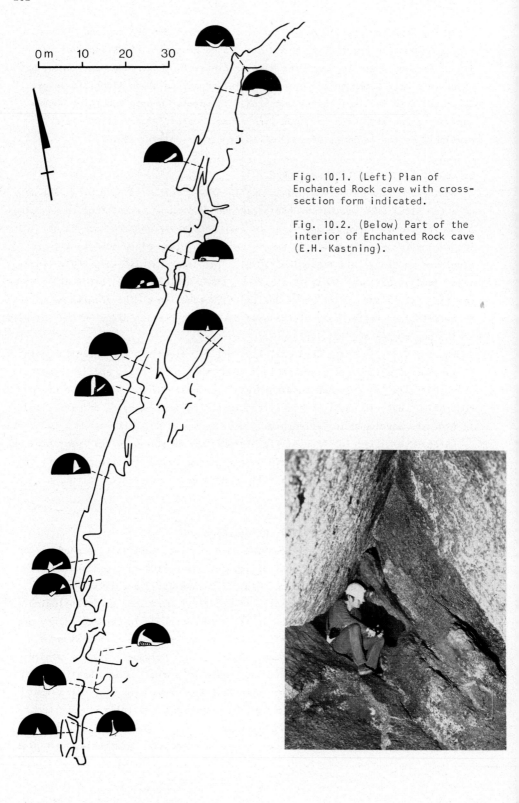

0 m   10   20   30

Fig. 10.1. (Left) Plan of
Enchanted Rock cave with cross-
section form indicated.

Fig. 10.2. (Below) Part of the
interior of Enchanted Rock cave
(E.H. Kastning).

Fig. 10.3. (a) Above, examples of boulder tafoni at Mt Hall, western Eyre Peninsula, S.A., (b) below, at Lei Yue Mun, Hong Kong (H-L. Tschang).

284

Fig. 10.3. Tafoni (c) at Ucontitchie Hill, Eyre Peninsula, S.A., (d) Remarkable
Rocks, Kangaroo Island, S.A.

Fig. 10.3. (e) Sheet tafoni, Pearson Island, S.A.

Fig. 10.4. Tafoni due in part to preferential weathering of joint intersections, northeastern Sardinia (R.F. Peel).

Fig. 10.5. Shelters associated with more fissile beds or scarp foot soil moisture attack acting on gritstone, English Pennines, near Addingham, West Yorkshire (Geol. Sur. Mus.).

widely developed (e.g., Fig. 10.5 - see also Fig. 3.22). Very small hollows are called *alveoles* (hence alveolar weathering - see Figs. 10.6 and 10.7), though some authors refer to them as honeycomb weathering (e.g. Demek, 1964a). The ceilings and walls of some tafoni are ribbed (Fig. 10.8 (a)), and others are inversely mamillated (Fig. 10.8 (b)), suggesting that alveolar development may be an early stage of tafoni formation. At many sites the inner walls and ceilings are characterised by 'books' of thin flakes which follow around the irregular mamillated surface (Fig. 10.9). Elsewhere, the granite is loose and disaggregated, but some walls are stable and apparently no longer subject to active disintegration. Large caverns are called *grottes*, or, in Australia, *shelters*, though these are usually located at the bases of cliffs, are called *cliff-foot caves*, and some lack significant visors. Cliff-foot caves are well developed in limestone (e.g. Lehmann, 1954; Jennings, 1976; Smith, 1978) and in sandstone (e.g. Peel, 1939, 1941; Twidale, 1978b).

In the context of granite bedrock, tafoni are developed on the undersides of sheet structure and of boulders, though they also occur on the sides of steep rock walls, hence, sheet tafoni, boulder tafoni (the tortoiseshell rocks of Jutson (1914) and other authors) and sidewall tafoni (Fig. 10.3).

Tafoni vary in size from a few centimetres radius to large hollows that are

Fig. 10.6. (a) and (b) Alveoles in various stages of development in central Namibia (B.B. Hambleton-Jones).

Fig. 10.6. (c) Alveoles in late stage of development in central Namibia (B.B. Hambleton-Jones).

metres across and high and in which a group of people can readily stand. They have been described from several climatic contexts, both interior and coastal (see e.g. Cailleux, 1953; Klaer, 1957; Calkin and Cailleux, 1962; Dragovich, 1964, 1966, 1969): Antarctica, Hong Kong, Corsica, Sardinia, central and south-ern Australia, and so on (Fig. 10.3). They do not appear to be developed on the Dartmoor granite, for example, but the famous split mushroom rock known as the Peyro Clabado (Fig. 4.9 (d)), in the Sidobre of southern France, is hollow on its underside and Demek (1964a) describes small tafoni and alveoles, known locally as dew holes, from the Bohemian Massif. Tafoni are rare or absent in the humid tropics.

(ii) Origin

The origin of tafoni can be considered under three headings: initiation, growth or development, and the nature of the visor.

(a) Initiation: There is no doubt that tafoni are present and are, indeed, initiated by soil moisture attack beneath the land surface. Boyé and Fritsch (1973) report hollows in the fresh rock surface occupied by grus and already present on the underside of sheet structure in a newly exposed quarry face at Ebaka, in southern Cameroon. Small hollows, formed by the evacuation under gravity, of especially intensely weathered rock at the intersection of joints, and along joint planes, have also been noted in a quarry face on northwestern Eyre Peninsula (Twidale and Bourne, 1975a).

Fig. 10.7. Alveoles in granite (a) above, Antarctica (R.L. Oliver) (b) at
Cheung Chau, Hong Kong (H-L. Tschang).

Fig. 10.8. (a) Mamillated ceiling of tafone, Ucontitchie Hill, Eyre Peninsula, S.A. (b) ribbed and scalloped ceiling, Tcharkuldu Hill, Eyre Peninsula, S.A.

Many cylindrical gnammas, or weather pits, (Twidale and Bourne, 1978b - see Chapter 8) widen at their bases, where they terminate in a sheeting joint. At John K Rock and Myrtle Rock, both on northwestern Eyre Peninsula, for example, the fractures have been enlarged by the weathering out of both the underside of the upper slab and of the upper surface of the lower (Fig. 10.3). The hollows so formed are incipient sheet tafoni. Two such forms have been located high on the western side of Ucontitchie Hill, in the walls of an irregular cylindrical hollow that intersects two sheeting joints. The merging of such an extended sheet tafoni and of a pan on the crest of Little Wudinna has led to the under-mining of the depression and the collapse of some of its overhanging perimeter wall.

In addition to such direct evidence, tafoni and flared slopes commonly merge with one another (Fig. 9.15). Examples have been noted on the western foot-slope of Ayers Rock (in arkose); in the sidewalls of a joint cleft on Scholz Rock, northwestern Eyre Peninsula and on Bloedkoppie Dome in central Namibia (B.B. Hambleton-Jones, pers. comm.). They are found also as isolated boulders on Kokerbin Hill, in the southwest of Western Australia and at Murphy's Haystacks, northwestern Eyre Peninsula (Fig. 10.10). Pronounced hollows are associated with basally fretted slopes (see Chapter 8) at several sites on Eyre Peninsula (Fig. 9.16) and in the Pietersburg area of the northern Transvaal (Fig. 10.11).

Alternatively, some workers have attributed tafoni to preferential weathering, arguing that the hollows were originally occupied by material that was different from, and presumably weaker than, that which remains (Reusch, 1883; Choffat, 1895-6; de Lapparent, 1907; Miller, 1940). It is difficult to test this hypothesis because the alleged different material has been eroded, so that comparisons are impossible. It would be strange, however, if weathering and erosion had everwhere proceeded to a stage where the different material has been completely evacuated. On the other hand it is clear that granites, like other igneous and metamorphic crystalline rocks, display significant primary petrological variations, both textural and compositional, and patches of such mafic minerals as biotite may well have been weathered more rapidly than the mass of the rock, producing a small initial depression or hollow. Such prefer-ential weathering, however, can in no way account either for the enormous tafoni observed in the field or for their preferred development on the undersides of blocks and sheet structure, or for their common occurrence in the scarp foot zone.

Other workers have suggested that concavities developed in the rock surface subsequently evolve into tafoni (Kvelberg and Popoff, 1937). The concavities are said to be either structural in origin and related, for example, to curved

Fig. 10.9. Thin books of flakes in ceiling of tafone, Ucontitchie Hill, Eyre Peninsula, S.A. Each flake is about 1 mm thick.

joints, or to result from uneven scaling (Wilhelmy, 1958). Again, however, the suggestion provides no explanation of the development of the forms or for their distribution.

(b) Development - salt crystallisation: Some workers have suggested that temperature variations within tafoni are sufficient to cause disintegration of the rock exposed in the walls and ceilings of the hollows and so bring about the enlargement of the negative forms (see Walther, 1900; Klaer, 1956). Others have argued that the microclimates found in tafoni are more stable with a smaller range of temperature and humidity than is experienced in the open air. From this it has been suggested (Dragovich, 1967) that hydration is more marked on the interior surfaces of tafoni than on the external walls of the rock masses, and that the minerals so affected cause fracturing and flaking of the rock. Granular disintegration is attributed by some to similar processes. Unfortunately, the rock forming the flakes and granules is essentially fresh and frequently displays little or no sign of alteration.

Many workers attribute the development of tafoni to flaking or granular disintegration due to salt crystallisation.

Theoretically, salts can disrupt a rock by crystal growth, by hydration expansion, by thermal expansion and by osmotic pressure (see Becker and Day, 1905, 1916; Birot, 1954; Evans, 1969; Winkler, 1975). Though thermal expansion is favoured by several workers, notably Cooke and Smalley (1968; but see also Goudie, 1974), it has been shown that temperature variations within tafoni

Fig. 10.10. Flared boulder with shallow tafoni, Kokerbin Hill, southwestern W.A.

Fig. 10.11. Small tafoni associated with basally fretted slope, Pietersburg, northern Transvaal, R.S.A. Hammer provides scale (M.L. Hugo).

are very limited (Dragovich, 1966, 1967). Hydration expansion, even in conditions of suitable temperature and relative humidity, is rejected by Kwaad (1970) as acting too slowly to be effective. Moreover, gypsum, the only important hydrated mineral commonly found in and near tafoni, is stable; fine-grained clays are scarce in tafoni walls. Thus, although some workers (White, 1973; Winkler, 1975) give weight to the process, it is likely to be of only minor significance overall. Osmotic pressure effects require some form of suitable membrane to be effective, and an essentially closed system, neither of which conditions is found in the field. Thus, crystal growth remains as the most likely effective mechanism involving salts, as was described more than a century ago by Thomson:

> ... the tendency of crystals to increase in size when in contact with a liquid tending to deposit the same crystalline substance must push out of their way the porous walls of the cavities in which they are contained. (1863, p. 35).

Many workers have in various degrees favoured the salt crystallisation mechanism (e.g. Mortensen, 1933; Bourcart, 1957; Winkler, 1965; Frenzel, 1965; Bruckner, 1966; Evans, 1969; Johnston, 1973; Goudie, 1971), though there are many variations on the theme. Rondeau (1961, pp. 170-173), for example, believes that salt contributes both mechanically and chemically to rock disintegration, acting through expansion or crystallisation, and through the ionising action of slowly evaporating water. Cooke and Smalley (1968) consider that insolation causes salts to expand more rapidly than the host rock - and so on; there are many variations on the salt theme, including some who assign to it merely a contributory or secondary role (see e.g. Dragovich, 1968).

Experimental work (see e.g. Thury, 1828) and observations of building stones that have crumbled as a result of coming into contact with natural or industrial solutions (see e.g. Thompson, 1863) have together convinced many researchers of the efficacy of salt crystallisation as a mechanism of rock weathering. And these two approaches continue to figure significantly in the task of convincing field scientists of the reality of salt crystallisation. Laurie (1925), Plenderleith (1956), Cole (1959), Winkler (1965), Roberts and Kallend (1978), and Hutton (1981) are just a few of the many who have concerned themselves with systematic studies of building stones, the disintegration of which is constantly in the public consciousness due to the reported weakening of the fabric of such well-known structures as the Taj Mahal and the Parthenon as a result of industrial air pollution. As for experimental work, Bakker et al. (1968) and Kwaad (1970) are merely amongst the most recent of a long line of scientists who have endeavoured to demonstrate the effectiveness of salt crystallisation in rupturing rocks in laboratory conditions (for review, see Evans, 1969). The mechanism has been given theoretical respectability by such workers as Andrée (1912), Taber (1966) and particularly Buckley (1951).

Salt crystallisation has long found favour with field geologists and geomorphologists. Cooke (1981) has recently provided a perceptive and balanced account, but long ago Jutson (1914) referred to exsudation, various German workers have described *Salzsprengung* and Klaer (1973) has strongly argued the case for salt-induced disintegration in Corsica. For obvious reasons, the process has been invoked in explanation of weathering in general, and of hollows called alveoles or tafoni in particular, in coastal and arid environments, for it is there that salts are most readily available. Bartrum (1936) and Tricart (1962) are two of many who have invoked salt crystallisation in explanation of honeycomb and similar weathering forms in coastal contexts, while Walther (1900), Wilhelmy (1964) and Goudie (1971), as well as Jutson, have resorted to the process in hot desert regions. Salt crystallisation in polar regions, particularly Antarctica, has been cited by many writers, including Cailleux (1962), Bardin (1963) and especially Wellman and Wilson (1965). Others, like Kelly and Zumberge (1961), concede that salt crystallisation may be responsible for the weathering forms they observed in Antarctica, but regard the mechanism as not proven.

The origin or source of the salts said to have crystallised and caused rocks to rupture has been, and remains, a matter of concern. The association of tafoni with coastal environments has encouraged the suggestion that the salts are of marine origin, and that they are carried on to the rocks either by waves or in fogs. Thus Hambleton-Jones (1976, p. 50) attributes both alveoles and tafoni to uneven granular disintegration due to salt crystallisation, the salt having been introduced in sea fogs that are common in that part of western Namibia in which he worked. Other workers invoke the crystallisation of salts derived directly from the sea (e.g. Högbom, 1912; Scherber, 1932; Denaeyer, 1956). Some regard the salts as cyclic or transported by the wind; hence, it is said, the profuse development of tafoni in deserts, both hot and cold. Salt is certainly transported in this manner. Clouds of salt have been seen being blown from the surfaces of salines by strong winds in several parts of Australia. The difficulty, both with marine and aeolian salts, or indeed salts carried in rivers or groundwater, is to explain how they can penetrate deep into rocks of low permeability. Some could infiltrate along fractures (hence those tafoni associated with joints), but many tafoni are associated with plane rock surfaces.

Connate salts, or salts derived from the rock itself, present no such problem. Most granites contain radicals of Ca, Na, SO_4 and Cl in their feldspars and micas. These are released on weathering, and combine and crystallise as halite or gypsum, exerting enough pressure to rupture thin particles of rock as they do so. This notion has been supported by Logan (1960) and Bradley *et al.* (1978) (see also Winkler, 1979; Bradley *et al.*, 1979). In order for the salts to be translocated within the weathered outer shell of the rock, however, water is needed. On the exposed upper surface and sides of boulders, for example, leaching of salts by rain is likely. Moreover, on such exposed surfaces, lichens and mosses grow and absorb water. But, it is on sheltered undersurfaces that tafoni develop for the most part. Here, dew may form, or the vapour respired by small mammals that find shelter in the crevices and hollows may be condensed.

The crystallisation of connate salts offers a plausible explanation of the flaking and granular disintegration observed to be actively taking place on the inner sidewalls and ceilings of tafoni. It is certainly more feasible than the disintegration due to heating and cooling favoured by some workers (Kvelberg and Popoff, 1937; Cailleux, 1953; Matschinski, 1954), for the range of temperatures within the hollows is only limited (Dragovich, 1966).

The distribution of salts in the rock, flakes and visors of South Australian tafoni is consistent with the laboratory experiments demonstrating the effect of crystal growth. The granites in South Australia have been exposed for a very long time to wet winters and hot, dry summers, so that even a slow rate of crystal growth from concentrated salt solutions could have produced large tafoni.

On the other hand, the question is clearly more complex. Fuge (1979), for instance, has cited the presence of chlorine in several British granites, and reported that of the areas examined chlorine attains its highest concentration in the granites of southwestern England. Yet neither there nor elsewhere in Britain are true tafoni in evidence. Again, although tafoni are widespread in the Upper South East and Eyre Peninsula regions of South Australia, not all boulders exhibit tafoni, and it is also difficult to understand how salt in concentration no greater than 0.5% can create 10-25% fracture porosity in otherwise fresh rock.

The probable existence of residual stress and its associated strain energy within these rocks (Kieslinger, 1960) may go some way to help explain this problem. Residual stress in granitic rocks can be produced in a variety of ways: cooling, erosional unloading, tectonism, and possible deuteric altera-tion. Stressed rock dissolves more readily than unstressed rock (Turner and Verhoogen, 1960, p. 476) and other weathering processes may also be facilitated by the presence of stress. This may help explain how such small amounts of salts (and water) can produce such dramatic results, though other factors are obviously involved in the formation of tafoni.

But, despite the theoretical, experimental and observational support and evidence, there are still problems concerning, in particular, the source of the salts that achieve disintegration of rocks during their crystallisation. Some consider them to be introduced into the system in sea water or in sea spray. In desert regions salts may be carried on the wind. Even so, some of the rocks, of which granite is a prime example, evidently affected by salt crystallisation are of a very low porosity and permeability, and it is difficult to understand how saline solutions could penetrate the rock mass, particularly in sheltered sites. It has recently been suggested that the salts originate in the rock itself, being derived from the products of earlier weathering (Hutton *et al.*, 1978).

The precise nature of the mechanism is also a matter of debate. For some, the force exerted by crystallisation is sufficient. Others (e.g. Cooke and Smalley, 1968), argue that as most salts have a higher coefficient of thermal expansion than have most rocks, this is enough to cause disaggregation. Rondeau (1961, pp. 170-173), on the other hand, invokes the ionising action of evaporating waters.

The question of salt crystallisation and its possible geomorphological consequences is complex. For the present it can be taken that salts are locally active, that their efficacy in weathering varies with local environment and with structure, and that salt crystallisation is one of several viable mechanisms of physical weathering that facilitate the readier ingress of water and thus prepare the way for alteration of the rock.

(c) Visor: The *visor* is vital to the formation of tafoni, but is difficult to explain. It certainly possesses great strength. At Yarwondutta Rock, for instance, a hollowed block, weighing four to five tonnes, stands on points of the case-hardened exterior that, together, cover no more than 0.5 m^2.

Flaking and disintegration wear away and eventually breach the pendant slab from the inside, suggesting that whatever has protected the visor is to be found on the outer surface. For Wilhelmy (1964) the visor merely reflects an outer skin of rock that has been dried and thus hardened. Certainly, thin surficial zones, lacking any discernible alteration, but nevertheless clearly more resistant than those immediately below them, have been noted in the field, R. Dahl (1966), for instance, reporting such zones about 3 cm thick. But Wilhelmy also argued that crumbling occurs on shady aspects with the visor on the exposed or sunny side, and no such preferred aspect has been demonstrated, despite several plots of tafoni at various sites.

But whatever the details and disputes, case hardening is obviously a factor in tafoni development and therefore deserves further consideration.

(iii) Case hardening and other coatings

Case hardening takes the form of a red-brown coating concentrated at the surface, but extending several crystals deep beneath it, that is commonly developed on exposed surfaces in arid and semiarid lands. It is associated with projecting lips, is obviously more resistant than the unaltered rock, and, for this reason, is widely referred to as case hardening. Although some writers use the term almost synonymously, case hardening differs from desert varnish, which term is used of a wide range of black, brown, yellow-brown and colourless patinas found in a wide range of climatic conditions, but particularly in the arid and semiarid tropics.

The brown, red, yellow and purple patinas are composed of silica and oxides of iron and manganese in various proportions (see e.g. Loudermilk, 1931; Engel and Sharp, 1958). Case hardening stands in contrast not only with true varnish but also with the black coatings frequently found in association with water (pools, rivers) in various climatic environments (see e.g. Humboldt, 1842, pp. 243-246; Francis, 1921; Twidale, 1980c) and, for this reason, known as 'river films' (Blackwelder, 1948). They are supposed to be of organic origin, and to consist of the remains of algae and lichen that have been concentrated in streams and wash (Francis, 1921).

The black encrustations so common in northern Australia and southern Africa, for example, may be of similar origin in part, though soot derived from seasonal (anthropogenic) burning may also contribute. Even so, the carbonaceous coatings are especially thick in gutters that score the bare rock surfaces. On Eyre Peninsula several of the gutters draining Mt Wudinna and Ucontitchie Hill carry encrustations of black material (Figs. 9.10 (a) and 9.29), and on Yarwondutta Rock and the Dinosaur some such grooves extend onto the overhanging slopes of the flared margins of the residuals. The black veneer is thin, discontinuous and relatively weak. It can, for example, be scraped away by sharp, needle-like leaves to give scratch circles (Fig. 10.12; see also Kukal and Adnan, 1970). But at Yarwondutta Rock and elsewhere it is notable that the coating has had a protective effect, for there has been localised relief inversion (Twidale, 1971a, p. 90; see also Fig. 9.21 (a)). In these instances, the flows of water originate in large measure in soil and detritus lodged in rock basins (see Chapter 8) or on ledges, so that they may well be flushing out dead organic material. Similar coatings occur on the flanks of Ayers Rock.

Turning to the origin of case hardening, many of the comments made in respect of varnish apply equally to case hardening. Anderson (1931) suggested that the case hardened visors associated with tafoni in Idaho are stained with iron oxide, and Rondeau (1961, pp. 179-173) thought that visors were impregnated with limonite. Engel and Sharp (1958), however, demonstrated that rust-coloured case hardening developed in the arid American Southwest consists of oxides of iron, manganese, silica and alumina. Electron probe work shows that, though the patina is uniformly thin (it is nowhere more than 100 μ thick and mostly less), it consists of two layers: an inner coat consisting of SiO_2 and Al_2O_3, but with some iron and manganese, and an outer zone composed wholly of oxides of these last named metals (Hooke *et al.*, 1969). Similar studies carried out on varnish developed on an olivine basalt from Arizona have demonstrated a layered and

Fig. 10.12. Scratch circles due to arcuate movements of sharp, needle-like leaves of such plants as the yacca (*Xanthorrhea* spp.) and spinifex (*Triodia* spp.) being wafted on the wind, and marking the black patina developed on granite surface. This example is from the Keith area of the South East district of South Australia.

botryoidal structure, but, again, a dominance of iron and manganese (Perry and Adams, 1978). The same study distinguished optically opaque or dark layers rich in manganese and rich in CaO and red layers depleted in manganese but rich in oxides or iron, alumina, silica and potassium.

Some writers have suggested a biological origin for the varnish of arid lands. Scheffer *et al.* (1963), for instance, regard it as due to blue algae which have oxidised iron and other heavy metal ions and concentrated them in superficial oxidised skins on stones and other surfaces. Dorn and Oberlander (1981) attribute desert varnish to *Metallagenium*-like bacteria capable of concentrating ambient manganese, silicon and iron on rock surfaces.

Another possible explanation has been prompted by the observation that iron oxides (probably geothite and haematite) are concentrated at the weathering front in some profiles developed on granitic rocks. Some are very pronounced, some only faintly visible, but such concentrations are commonly present. Below, the rock is intrinsically fresh and cohesive; above, the thin iron oxide zone is bleached and altered, presumably as a result of the leaching and illuviation

of soluble salts from the surface to the weathering front, where they are concentrated, the solutions being able to penetrate no further into the impermeable
rock. This bleached, outer zone, which varies in thickness between 2-3 mm to
4-5 cm, remains hard when dry but becomes soft and friable when wet.

It is suggested that granite masses, whether boulders or inselbergs,
initiated in the subsurface by differential weathering, may acquire such a
marginal concentration of iron oxides; that the weathered, outer, bleached zone
is eroded after exposure; and that in this way the fresh rock masses come to
have a coating or patina rich in iron which is enough to protect the underlying
rock against epigene weathering and thus allow a visor to develop. (This
suggestion is being further investigated, both in the field and in the
laboratory.)

Thus tafoni appear to be initiated by moisture attack, either in sites of
structural (mineralogical, textural, fracture intersections) weakness, in the
moist scarp foot zone, or on the shady and moist undersides of blocks and slabs.
Subsequent development may be due to salt crystallisation and associated rock
shattering in the form of either granular disintegration or flaking; and visor
development most likely due to minor accumulations of oxides of iron, manganese
and possibly silicon on the external walls of blocks and slabs.

Tafoni are not restricted to any single, conventionally defined climatic
region, being well developed in coastal contexts in various climates, and, in
any case, being represented in various arid and semiarid climates, hot and cold.
But they do not occur in humid tropical regions, possibly because, the frequent
rains and the ready growth of mosses prevents the accumulation of any form of
varnish or crust, so that a visor cannot develop. Similarly, in nival (as
opposed to glacial) regions, and in temperate lands subject to nival action now
or in the recent past, freeze-thaw may have prevented the formation of a visor
and, hence, of true tafoni.

CHAPTER 11

SPLIT AND CRACKED BLOCKS AND PLATES

Many granite blocks, boulders, slabs and plates have been split as a result
of the development of fractures, either single or arranged in distinct, if
irregular and varied, patterns.

A. SPLIT ROCKS

Some boulders have been split into two parts (Fig. 11.1). In many instances
the two parts are equal, or nearly so. The parting fractures are most commonly
planar, but some are arcuate or conchoidal and others of a distorted sigmoidal
shape in section. Some *split rocks* develop beneath the land surface (Ollier,
1971) but others are clearly of epigene origin.

Fig. 11.1. Split boulder at the Devil's Marbles, central Australia, with, at
right, a boulder with a secondary fracture.

Several writers have attributed split rocks to heating and cooling under hot desert conditions (e.g. Hume, 1925, pp. 11-38; Hills, 1975, pp. 132-133). But bearing in mind the large volume of rock involved, the essentially superficial nature of insolation changes, and the poor conductivity of rocks, it seems doubtful whether heating and cooling alone, even aided by rain showers, could achieve the splitting of large isotropic masses, though cobbles and pebbles (see, for example, Hume, 1925, Pl. IV) are evidently so affected. On the other hand, several writers believe that rapid cooling of heated blocks provides a reasonable explanation for split rocks, and Whitaker (1974) records the splitting of rock near Halls Creek, in northwestern Australia, in 1952 shortly after a rainstorm. Unfortunately, it is not known whether the split developed along a pre-existing latent joint. In any case, this mechanism cannot be of general application, for split rocks occur in such humid tropical regions as the Tampin area of West Malaysia (Fig. 11.2), in temperate regions like the Mount Lofty Ranges and Eyre Peninsula, and in nival areas like the Pyrenees and the Kosciusko region of southern N.S.W. Some writers, like Ollier (1971), attribute split rocks to pressure release, but the geometry of the split blocks, their size and likely accumulated stress, and the general history of boulder development render this suggestion unlikely. The most plausible general explanation of split rocks is that secondary fractures (i.e. fractures not involved in delimiting the joint blocks) within the original mass have been exploited by moisture attack (most

Fig. 11.2. Split boulder near Tampin, West Malaysia; X marks the parting.

likely hydration shattering) either in the subsurface or after exposure (Louis, 1960, pp. 38-39; Twidale, 1968a, pp. 100-104). Though the boulders and blocks are defined and delineated by orthogonal joints, it is a matter of observation that they include other secondary or latent joints (Fig. 11.3). These subsidiary fractures are penetrated by moisture and so widened, either as a result

Fig. 11.3. Stages in the development of a split boulder (a) secondary parting in boulder at the Devil's Marbles, (b) block split by fracture, but with the two parts still closely justaposed, in the Margeride of the southern Massif Central, France.

Fig. 11.3. (c) Third stage in the development of a split boulder, at Palmer, with the two parts just separated.

of freeze-thaw activity (as in the Pyrenees) or in consequence of alteration of the rock immediately adjacent to the parting. Weathering along the secondary fracture weakens cohesion between the two adjacent masses and, unless the blocks have flat bases and rest on an even platform, their weight causes them to fall apart (Figs. 11.3 and 11.4). Certainly when the fractured block occurs on a slope (Fig. 11.5), separation has occurred. On the other hand, some blocks and

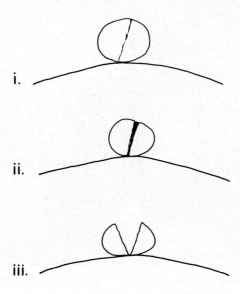

Fig. 11.4. Suggested mode of development of split rock.

Fig. 11.5. Split and separated blocks, with secondary fracture visible in upper block, Kokerbin Hill, W.A.

boulders have been split twice and thrice. For these, frost riving affecting secondary, latent fractures offers the most satisfactory explanation, for all occur in areas subject to frequent freeze-thaw alternations, either at present or in the recent past. Even so, many sit on a foundation too even and stable for the separate blocks to fall apart.

B. PARTED BLOCKS

Some *Kluftkarren* form complex patterns and some of the blocks defined by them have apparently been joggled or moved. Such *parted blocks* consist of two cubic or quadrangular blocks that were, presumably, closely juxtaposed originally, but which are now separated by a considerable gap (Fig. 11.6). In some instances, both gap and marginal blocks are surmounted by other blocks (Fig. 11.7).

These forms cannot be accounted for by unbuttressing and the two adjacent blocks rotating outward in opposed directions, as suggested in explanation of split boulders. The blocks in question are angular and could not roll apart, even if the blocks to either side were eliminated. In some instances, the parted blocks are surmounted by other large residuals, the weight of which would surely prevent the underlying masses from moving, save in the most unusual

Fig. 11.6. Simple parted block at Bellever Tor, Dartmoor, developed in granite with well developed flaggy structure or pseudobedding.

circumstances (e.g. as in earthquakes).

The partings are not due to the weathering and erosion of sills of weaker rock, for no sign of such materials is found in association with them. They cannot be attributed to the slippage of blocks under gravity, for, as Worth (1953, pp. 25-26) pointed out, the parted blocks characteristically stand on sensibly horizontal bases. Nor can massive frost wedging be cited as an explanation of all parted blocks, for, though it could conceivably have operated on Dartmoor, for instance, during the Pleistocene, it can safely be ruled out for the low altitude tropical examples of parted blocks.

Instability and creep consequent on seismic shocks provides another possible mechanism, but not all the areas whence parted rocks have been reported are especially unstable.

Nor is it possible to argue that scarp foot weathering could so undermine the outer areas of the blocks that they would tilt outwards. There is no consistent evidence of such undermining, though it occurs in places. The superincumbent blocks has a stabilising influence, and the inherent strength of the rocks can accommodate such undermining without collapse taking place. Thus there is at present no satisfactory explanation of parted blocks.

C. POLYGONAL CRACKING

Polygonal cracking or weathering consists of a pattern of shallow fractures that is typically polygonal, though square and rhomboidal shapes also occur (Fig. 11.8). Isolated patches of superficial fractures, called heiroglyphs, may be related, or incipient, forms (Fig. 11.9). Examples have been recorded from

Fig. 11.7. Complex parted blocks with upper blocks located over gaping voids
(a) above, at the Devil's Marbles, N.T., and (b) below, at Bellever Tor,
Dartmoor; note flaggy structure.

various parts of the U.S.A. and of Australia, southern Africa and elsewhere.

The pattern was discussed, albeit rather inconclusively, more than half a
century ago (Johnson, 1927; Leonard, 1929), since which time it has received
little attention, though it has been noted or illustrated incidentally (e.g. in
Blackwelder, 1929, his Fig. 2 at p. 395; Wilhelmy, 1958, p. 170).

(i) Description

Most of the examples discussed here are developed on granite bedrock on Eyre
Peninsula and in the Pilbara of Western Australia. Polygonal cracking is not

restricted to vertical faces, as claimed by Leonard (1929), but occurs also on sloping, and even overhanging, faces (on the undersides of boulders, for example), as well as on essentially flat surfaces. Polygonal cracking occurs on granite boulders at various erosional levels (see Twidale and Bourne, 1975b) on Tcharkulda Hill, near Minnipa, on northwestern Eyre Peninsula, where it is found in great profusion; on Wallala Hill and Corrobinnie Hill, also on Eyre Peninsula; just west of Broken Hill, in western N.S.W.; near Naraku, north of Cloncurry, northwest of Queensland; and especially in parts of the Pilbara of Western Australia.

Fig. 11.8. Polygonal cracks on boulder at Tcharkuldu Hill, northwestern Eyre Peninsula, S.A.

It is also developed on quartzite boulders in the lower Mount Arden Valley, some 27 km NNE of Quorn, in the southern Flinders Ranges. On Tcharkuldu, Wallala and Corrobinnie hills (Fig. 11.10) the cracking is well developed on high-level erosional granite platforms. The fractures that constitute the polygonal patterns are evidently initiated as fractures propagated from specific points, for on boulders at Corrobinnie Hill and elsewhere star fractures are developed (Fig. 11.11). They spread and merge with adjacent cracks to form polygonal patterns. They differ in several respects from those that comprise the blocky disintegration that has been described from the quartzitic Mt Conner, in central Australia (Ollier and Tuddenham, 1962; Ollier, 1963), and which also occurs on quartzite boulders in the same area of the lower Mount Arden Valley where polygonal cracking is also developed.

Fig. 11.9. Irregular and discontinuous superficial fractures, called heiroglyphs, on boulder at Daadening Hill, near Merredin, W.A.

As the name implies, the fractures involved in polygonal cracking delineate regular geometrical forms, extend to only shallow depths and terminate in fractures that run parallel to the local surface, whereas those involved in blocky disintegration are irregular. The fractures delineating the polygonal plates are normal to the surface, whereas those involved in blocky disintegration display various orientations.

Most of the fractures involved in polygonal cracking are narrow and become narrower in depth. Some have been made wider as a result of weathering, some observed by Leonard (1929) being 5 cm across. The juvenile narrow fractures

Fig. 11.10. Platform with polygonal cracking, Corrobinnie Hill, northern Eyre Peninsula, S.A.

are linear and the edges cut through individual crystals, but those widened by weathering are in detail jagged due to the projection of individual crystals, particularly of quartz.

The polygonal plates range in diameter from 2 cm to some 24 cm, with the average and mode both near the upper end of the range. Some of them are slightly curved or convex upward in respect to the surface of the host rock (cf. Leonard, 1929, p. 491). They are delineated at a depth of a few centimetres by fractures disposed parallel to the rock surface, so that, in effect, the cracking is confined to thin plates. On boulders on Tcharkuldu Hill and near Mt Magnet, Western Australia, some of the thin plates have been either worn away or have fallen, so that as many as three layers, each with polygonal cracking developed, are exposed (Fig. 11.12). The narrow hairline cracks of the deeper layer stand in marked contrast with the wider, less sharply defined, fractures of the outer skin.

As Johnson (1927) noted also in Maine, the granitic rocks in which polygonal cracking occurs are characteristically even-grained. Thus, the granite at Tcharkuldu Hill, where the cracking is most abundantly developed, is of this type. Conversely, where the granite is porphyritic, as, for instance, at Mt

Fig. 11.11. Star fracture on boulder at Corrobinnie Hill, northern Eyre Peninsula, S.A.

Wudinna, no polygonal cracking has been observed.

(ii) Origin

(a) Previous work: Johnson (1927) attributed polygonal weathering, as he called the cracking, to the development of isolated systems of fracture through insolation and chemical weathering and to their subsequent propagation and inter-section. This suggestion is difficult to sustain in light of the freshness of the rock adjacent to the first juvenile fractures, and the even distribution of the cracking. If insolation were responsible, either wholly or in part, then on Eyre Peninsula, for example, and all else being equal, the cracking ought to be better and more commonly developed on the northern and western faces of the boulders than on the southern and eastern; more on the upper than on the under side of the forms. A survey of the polygonal cracking on Tcharkuldu Hill, where there are sufficient specimens to constitute a significant population, shows

Fig. 11.12. Multiple shells, each with polygonal cracking, near Mt Magnet, W.A.

that the cracking is, to all intents and purposes, evenly distributed with respect to aspect, though there is more on upper surfaces than on the undersides of boulders. Nevertheless, that there is any cracking on the shady undersides is significant.

Leonard (1929) related the cracking to the history of the rock in which it is developed. According to Leonard, polygonal cracking is a primary feature 'developed on the joint faces at the time of major jointing and in consequence of it' (Leonard, 1929, p. 491). This explanation implies that the surfaces on which the cracking occurs are all joints; and this is not so. Cracking has been observed on only one undoubted joint face, at Naraku, and that is pitted, indicating that it has suffered some modification by weathering (Twidale and Bourne, 1976a).

Almost everywhere, the cracking is found on surfaces shaped by weathering. Most of the boulders on which the polygonal cracking occurs are one-time corestones developed by joint-controlled differential subsurface weathering (Chapter 4) and subsequently exposed by erosion. Others, and particularly those on Tcharkuldu Hill, are most likely derived from the disintegration of sheet structures, but weathering of the joints and rounding of the blocks is implied. Again, then, the surfaces are not of structural origin but are due to weathering and erosion.

Structures developed in the original magma may have influenced the spherical shape of some corestones and boulders (see Twidale, 1968a, pp. 97, 101, 103; Twidale, 1971a, pp. 31-32; and Chapter 4), but even in these, possibly rare, cases, the shapes are exposed as a result of weathering.

In passing, it is interesting to note that Leonard (1929) dismissed weathering as a possible causation because he thought it was inconsistent with the observed facts, and argued that if the polygonal cracking were due to weathering processes the faces of the plates so produced should be bulging or curving. He was able to find no such evidence, but, as described earlier, some of the plates on Tcharkuldu Hill stand a few millimetres higher in the centre than at the edges.

(b) Analysis: Polygonal cracking is a result of a disequilibrium at the outermost shell or shells of rock, and could have been caused by the contraction of the surface layer or layers in consequence of either weathering and/or volume reduction; the contraction of the core of the boulders and the collapse of the outer shells on to the reduced core; the expansion of the core causing the outer shell or shells to fracture in adjustment to this internal pressure; the expansion of the outer shell for some reason other than internal pressure.

Alteration is not an essential feature of or prerequisite for the development of cracking. The rock bordering juvenile cracks is quite fresh. The fractures allow water to penetrate through the outer shells, and are the foci of weathering activity, but the cracks are not a result of these processes. Thus, contraction of the outer skin by weathering is ruled out.

If it is accepted that many boulders are weathering forms, it is difficult to see why contraction should take place at such a late stage in the history of the rock. The plates show no sign of overriding at the edges in consequence of fitting to a smaller core; on the contrary, there is a tendency for the cracks to gape, though this results partly from weathering after their formation. For these reasons, contraction of the core is precluded.

Similarly, it is difficult to visualise why expansion of the core should occur after the boulders have been reduced and shaped by weathering and erosion. In any case, any such radial expansion would surely be accommodated along crystal cleavage, etc. Farmin (1937) has invoked such radial expansion in explanation of onion weathering - the development of concentric shells, each of the order of 10-20 cm thick, in granite boulders. These fractures are significant in the present context, but they cannot be due to expansive stress (Wolters, 1969; Ingles et al., 1972). If expansion were the reason for their development they ought, surely, to be best developed on those parts of the boulder subject to least confining pressure, i.e., on the upper slopes and on

the upper sides of masses. There is no suggestion of marked preferential development. Polygonal cracking is well developed on platforms that are quite
extensive, being of the order of tens of square metres in area, and cut across
several individual joint blocks. There is no suggestion of a single centre of
expansive stress from which cracking has been propagated.

Turning to the possibility of cracking being due to expansion of the outer
shell or shells, the latter, whatever their origin, are consistently associated
with polygonal cracking. Such partially detached shells could be cracked by
radial pressures, provided that their ends are fixed. Several possible causes
of radial stress are discussed.

Many fractures defining shells contain a mixture of mineral and organic
debris. The mineral debris consists of finely comminuted feldspar and quartz
fragments, with only a little clay. The organic material comprises leaf fragments, skeins of spiders' webs, moths' wings, and, most commonly, a tar-like
substance mixed with the other debris as discrete blobs and irregular skins.
Tests carried out to ascertain the pressure exerted by such materials on wetting
showed that the expansion initiated is inadequate to account for the fracture
of fresh crystalline granite.

Exposed rock surfaces attain very high temperatures as a result of insolational heating. Rock surface temperatures in excess of $50^{\circ}C$ have been recorded
in the Egyptian desert and even higher levels $(80^{\circ}C+)$ on soils in equatorial
Africa (see Cloudesley-Thompson, 1965), but there is no evidence that ordinary
heating and cooling cycles can themselves cause rocks to crack or shatter (see
e.g. Blackwelder, 1925; Griggs, 1936), but rapid cooling associated with desert
rain storms (see Hume, 1925, p. 27) may do so: it may be that surface layers or
individual minerals expand under the influence of insolational heating and that
the stresses caused by rapid cooling induce fracturing at one scale or another.

If such rapid cooling were responsible for polygonal cracking, however, the
fractures ought to be most commonly found on upper slopes of boulders and
inselbergs. They are, indeed, well developed on platforms at Corrobinnie Hill,
Tcharkuldu Hill and Wallala Hill, on Eyre Peninsula, but they are at least as
common and well developed on the sides of large residual boulders, and they are
also represented on the underslopes of these forms.

Expansion caused by the intense, if ephemeral, heat of bushfires could, in
theory, cause the expansion and cracking of the outer shells or rock (see
Blackwelder, 1927; Emery, 1944). According to Warth (1895) thick slabs of rock
arch from rock masses when they are subjected to intense heating, and it is
conceivable that with thin slabs the radial stresses so introduced could result
in the development of tangential fractures.

The field evidence argues against such intense heating playing any part in
the development of polygonal cracking. Discontinuous flakes of rock, rather than

shells, are produced by bushfires (see e.g. Figs. 3.4 and 3.5). Also, wide areas of northern Australia and of northern and southern Africa, for example, are deliberately and systematically burned every year or so, and if fires had any part in the development of polygonal cracks, the latter ought to be common in these areas. They are present (Fig. 11.13), but are not notably well represented. They have not been reported from formerly wooded areas such as Dartmoor, which might be expected to have experienced firing.

Again, there are reasons for suggesting that the cracking is initiated below the land surface (see below). If this is correct, bushfires, and any other form of heating, are ruled out.

(c) Possible explanation: Several observations and arguments suggest that polygonal cracking is initiated beneath the land surface. Polygonal cracking occurs on the inner shell of a corestone exposed in a road cutting in the Snowy Mountains, N.S.W. (Fig. 11.14). Cracking occurs on the inner shells of boulders, for example, near Mt Magnet, Western Australia. It is seen to be present when the outer shells are removed. Cracking is commonly associated with pitting and with case hardening, the first of which, certainly, the second, possibly, develops beneath the land surface. Well developed cracking is also associated with superficial concentrations of oxides of Fe and Mn. And finally, polygonal

Fig. 11.13.   Polygonal cracks on boulder near Neue Smitsdorp, in the northern Transvaal.

cracking is found on a low platform at the northern end of Tcharkuldu Hill, only a few metres away from boulders and grus *in situ*, indicating that the platform has only recently been exposed.

It has been suggested that the accumulation of carbonates as a consequence of pedogenic or weathering processes can create a space problem which is relieved by the arching of the near surface duricrust layers (Price, 1925; Jennings and Sweeting, 1961). In similar fashion, the weathering of the granite could have caused volume increase following the translocation of salts and their concentration in particular horizons within the weathered area (see e.g. Harriss and

Fig. 11.14. Several of the corestones set in grus and exposed in a road cutting in the Snowy Mountains, N.S.W., display polygonal cracks.

Adams, 1966; Wolff, 1967). Oxides of manganese and, to some extent, iron are
concentrated in some horizons so that they impregnate the rock mass. The Mt
Magnet area of central Western Australia and the Nanutarra region of the western
Pilbara come to mind. These concentrations of salts could have caused, first,
the development of concentric shells or rocks; second, with further weathering
and concentration of iron and manganese, tangential expansion and cracking of
the shells (Fig. 11.15); and third, the intersection of these star fractures
(Fig. 11.11) to give polygonal patterns. Such a mechanism is attested by the
upward arching of some of the plates defined by the polygonal cracks. On the
other hand, such concentrations of translocated salts are not found in association with all examples of polygonal cracking.

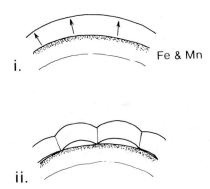

Fig. 11.15. Development of polygonal cracking as a result of precipitation of
oxides of Mn and Fe along concentric partings within boulders.

D. DISPLACED SLABS AND BLOCKS

   Some granite forms consist of slabs and blocks that have been split, cracked
and displaced. Several workers, reporting from varied climatic and lithological
environments, have described forms which are evidently of tectonic origin, and
are attributed to the release of compressive stress (Twidale and Sved, 1978).
By far the most common member of this suite is the A-tent. Others are overlapping slabs, displaced slabs, and vertical and horizontal wedges. The
'blisters' named by Blank (1951a) appear to be related forms. Other forms
consist of simple split and displaced blocks, and yet others of thin plates
defined by cracks that form polygonal patterns.

(i) A-tents

   A-tents consist of pairs of slabs, each roughly rectangular in shape, each
touching the adjacent rock surface at their outer extremities, but standing a
few centimetres above the general surface level where they are in contact in the
centre of the feature. They delineate triangular cavities, as do tents without
walls: hence their name (Jennings and Twidale, 1971; Jennings, 1978; Fig.

318

11.16). They are known as pop-ups in North America (see e.g. Coates, 1964; Sbar and Sykes, 1973). A-tents in granite are especially well developed in the Wudinna district of northwestern Eyre Peninsula, but they are fairly common throughout the region, and are known elsewhere. One, near Lithonia, Georgia, U.S.A., was featured by Dale (1923). They have been reported from Labrador (Peterson, 1975) and the Canadian Rockies, from Sabah, east Malaysia (Smith and Lawry, 1968), from the Rupununi savannas of Guyana (J.B. Bird, pers. comm.) and the Llano of central Texas (Blank, 1951a - Fig. 11.17). They have been described from near Comet Vale, Western Australia (Jutson, 1934, p. 254), from the Kulgera Hills, Northern Territory, the Broken Hill district of N.S.W. and the eastern Kimberleys of Western Australia (Jennings, 1978).

They are developed in sedimentary rocks on the Coolamon Plain (limestone), in marble at Wombeyan Caves, N.S.W. (Jennings, 1978), and in sandstone in Wyoming (Scott, 1897, p. 223; Pirsson and Schuchert, 1915, p. 20).

Fig. 11.16. An A-tent on the eastern midslope of Mt Wudinna, northwestern Eyre Peninsula, S.A. The slab X has slipped from immediately upslope into the cavity formed by the raised slabs.

Most of the A-tents that have been studied are 10-15 cm high and involve slabs about 10 cm thick, but one on Mt Wudinna has slabs up to 580 mm thick standing 820 mm above the floor (Fig. 11.18), (A_1 in Fig. 11.19), and one on Carappee Hill comprises slabs only 13 mm thick (Fig. 11.20).

A-tents are of two types. Angular tents comprise two planar slabs leaning

Fig. 11.17. A-tent on the lower southern midslope of Enchanted Rock, central Llano, central Texas.

against each other. The arched type is similar, but both the slabs are slightly curved convex upwards. The terminal and crestal fractures typical of the angular forms are not consistently present in the arched forms. Most angular tents occur on midslope sites, whereas arches are found both there and on the crests of hills. Arched slabs are fairly commonplace in tropical regions, but one at least occurs on Dartmoor, at Mistor (Worth, 1953, his Pl. 15A). At Cash Hill, western Eyre Peninsula, an A-tent has been exposed by the clearing of the regolith, suggesting that it developed beneath the land surface (Twidale and Bourne, 1975a).

Fig. 11.18. A-tent consisting of very thick slabs, and located on the western midslope of Mt Wudinna, S.A.

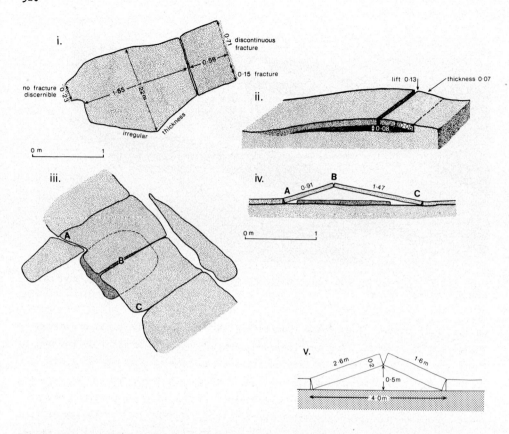

Fig. 11.19. Plan and Section of A-tents on Mt Wudinna (i-iv), and on Lightburn Rocks (v).   In (iii) slipped slab of Fig. 11.16 indicated.

Fig. 11.20. A-tent made of very thin plates of gneissic granite, Carappee Hill, northeastern Eyre Peninsula, S.A.

Measurements of A-tents from several sites on northern Eyre Peninsula suggest that, if, as seems certain, the presently raised slabs that constitute the A-tents were originally part of the smooth hillside, their present combined lengths would in most instances exceed the space they originally filled by some 3-4 percent (Jennings and Twidale, 1971; Twidale and Sved, 1978). One on Lightburn Rocks suggests an expansion of 5 percent (Fig. 11.19). But closer examination shows that though this is true of the thick slabs, fracture and uplift of the thin ones involves contraction (Twidale and Sved, 1978).

One of the A-tents on Mt Wudinna (Fig. 11.16, A_3 in Fig. 11.19) is floored by a slab of rock that has slipped a few centimetres downslope into the cavity formed by the two raised limbs of the A-tent.

(ii) Overlapping slabs

On the northwestern midslope of Mt Wudinna there is an *overlapping slab* consisting of a plate of granite the upper end of which is raised and overlaps another that forms the adjacent upslope section of the hillside by about 300 mm (Figs. 11.21 and 11.22). The two irregular edges of the raised and the flat slabs are disposed roughly normal to the surface and match perfectly, suggesting that they were once juxtaposed.

This overlapping slab cannot be explained as due to slippage, for the lower of the two components involved laps over the upper and not the reverse, as would be the case if downslope movement were involved. Furthermore, there are no gaps at either the lower or upper end of the two slabs involved. The feature could

Fig. 11.21. This overlapping slab on the upper midslope of Mt Wudinna cannot be attributed to slippage because the upslope member lies beneath its downslope counterpart.

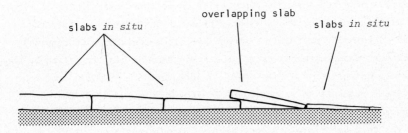

slabs *in situ*   overlapping slab   slabs *in situ*

Fig. 11.22. Section through overlapping slab shown in Fig. 11.21.

be a collapsed A-tent, but the collapse has not been caused by attrition of the
rock near the erstwhile crestal fracture, because the two opposed ends of the
slab match perfectly and the rock shows no signs of undue weathering. If an
A-tent collapsed with the downslope slab coming to rest on the upslope member,
it must have been caused by some catastrophic event.

A similar overlapping slab occurs on the northeastern midslope of nearby
Little Wudinna Hill. It now overlaps the adjacent slab by a matter of 15 cm,
and another platey fragment has slipped beneath the raised slab.

(iii) Displaced slabs

Displaced slabs occur elsewhere on Mt Wudinna, but perhaps the most spectacular example is seen on Little Wudinna Hill, where a triangular sheet some 410 mm
thick and 9 m long has slipped some 8.5 m down a 16° slope, coming to rest
against the steepened and flared lower slope of the hill (Figs. 11.23 and 11.24).
The triangular gap whence came the slab remains identifiable. Several more,
similar, slabs remain in a tumbled mass below the triangular one. Above the gap
left by the slabs, the sheet of rock, still *in situ*, is arched, though there is
no crestal fracture and hence no A-tent.

A few metres away to the west, a thick block, some 10 m long and involving
part of the flared slope, has slipped downslope a matter of 2 m. And on Mt
Wudinna, a slab roughly 5.4 m by 7.5 m and 320-530 mm thick has moved downslope
230-260 mm on a 19° incline (Fig. 11.25).

Displaced slabs are anomalous forms because there are many examples of blocks
and boulders perched precariously on slopes steeper than those on which slippage
of slabs has occurred. Thus, at Pildappa Rock there are large, sub-angular
joint blocks, more than one metre diameter, standing on slopes of 24° and 26°
(Fig. 11.26). Yet they are stable. The reason is that, though in general view
smooth, the granite slopes and the surface of the granite slabs and blocks
themselves are, in detail, very rough or pitted (Twidale and Bourne, 1976a -
see also Chapter 3). A micro-relief of a few millimetres has developed and,
even when lubricated by water, friction inhibits slippage on these slopes. Yet
some slabs have clearly migrated under gravity on slopes that are more gentle

Fig. 11.23. Triangular slab on the eastern side of Little Wudinna Hill, north-western Eyre Peninsula, S.A. The space whence it came is clearly visible, the point B originating at A.

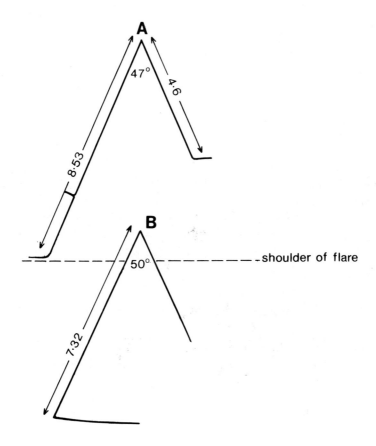

Fig. 11.24. Plan of triangular slab shown in Fig. 11.23.

324

Fig. 11.25. This slab on the northeastern upper slope of Mt Wudinna has moved
a few centimetres downslope, despite the relatively gentle incline and the
roughness of the granite surface.

Fig. 11.26. These large residual blocks are firmly perched on the upper
shoulders of Pildappa Rock, northwestern Eyre Peninsula, S.A. But the slopes
on which they stand are steeper than those from which other blocks and slabs
(see e.g. Figs. 11.23 and 11.25) have slipped.

than those on which blocks stand firm. One possible explanation is that tree roots have, during their growth and thickening, raised slabs of rock to such an extent that friction has been overcome. Vertical displacement of granite slabs has taken place at one site on Mt Wudinna (Fig. 11.27), but no lateral displacement is involved, and presumably the slabs will return to, or close to, their original position once the tree has died and rotted.

Fig. 11.27. During their growth the roots of this tree, on the upper slopes of Mt Wudinna, have displaced slabs, some of which weigh roughly half a tonne.

(iv) Wedges

Low on the western slope of Mt Wudinna is a triangular wedge of rock more than 13 m long and varying in breadth across the upper surface between 40 cm and more than 2 m (Figs. 11.28 and 11.29) which, in places, has been raised more than 150 mm above the level of the adjacent slope. A similar form has been observed at Enchanted Rock. These are *vertical wedges*.

Another small wedge, fractured and displaying systematic variation in the amount by which the constituent rock slabs have been raised, so giving the whole a dentate appearance, has been noted on the northeastern flank of Little Wudinna Hill, only a few metres from the overlapping slab previously noted.

Wedges of triangular cross-section are fairly commonly developed at the exposed lower edges of sheet structure. They differ from vertical wedges in

Fig. 11.28. This vertical wedge of triangular cross-section is located on the lower slopes of Mt Wudinna.

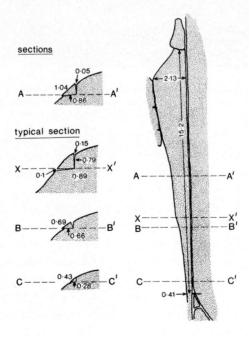

Fig. 11.29. Plan and sections through vertical wedge shown in Fig. 11.28.

that several have been displaced *laterally* so that despite seeming varied from the previously described forms, they, too, appear to be associated with the release of compressive stress (Chapters 2 and 5; Fig. 2.31).

(vi) Origin of the forms

Angular buckles affecting sandstone, and identical with the A-tents described here, have been illustrated from Wyoming, U.S.A. (Scott, 1897, p. 223; Pirsson and Schuchert, 1915, p.20) and attributed to insolation. But the sun's heat can be discounted as the cause of the A-tents and the other forms considered here, because insolation argues a reversible event, whereas the A-tent represents an irreversible change in the condition of the granite involved. Displacement by such a heating and cooling is conceivable, but permanent deformation is not. Furthermore, there is some suggestion that A-tents are initiated in the subsurface beyond the range of either insolation or the heat of bushfires (Twidale and Bourne, 1975a).

Insolation cannot account for the overlapping slabs, the wedges and the slipped slabs. Why, for example, should one limited area of the exposed outcrop be heated in preference to others so as to produce expansion? It is also extremely unlikely that heating could preferentially affect the lower, exposed parts of sheet structure to produce wedges, and particularly the lateral displacement of these forms.

There is no sign of significant chemical alteration having affected the slabs and blocks of granite involved in the forms, so that this need be considered no further.

The suggestion that A-tents are a manifestation of pressure release consequent on erosional offloading (Jennings, 1978; see also Peterson, 1975) does not account for the varied results of arching involving expansion in some places but contraction elsewhere. It does not explain why the angular tents are typically developed on midslope, and it is not consistent with the survival of the host masses of granite in inselbergs which have resisted weathering and erosion, because they are massive, monolithic and in compression (Twidale, 1971a, 1973). Had there been expansion, the vertical wedge on Mt Wudinna would surely have subsided rather than been thrust above the general level.

The other forms in the suite cannot be explained by such pressure release. A-tents and associated forms are manifestly youthful, yet occur on host forms that are evidently of some antiquity (Twidale and Bourne, 1975b), which is inconsistent, for if due to pressure release the minor suite would surely have formed as soon as the host forms were delineated by differential subsurface weathering long ago.

There is another line of evidence favouring compressive stress, and difficult to refute. Coates (1964) has reported that during the quarrying of Palaeozoic limestone in Ontario, the floor cracked and heaved, forming an A-tent the crest of which stood 2.4 m higher than it did originally, and affecting rocks some 15 m to either side of the crestal fracture. Calculations showed that pressure release could not quantitatively account for the upheaval, which was attributed to horizontal stress in the limestone.

Similar contemporary arching associated with rock bursts in a granite quarry in Tocumweal, N.S.W. is reported by Denham (in Jennings, 1978, *Addendum*; Denham *et al.*, 1979).

The forms described in this section are readily explained as being associated with the release of compressive stress, particularly if any remnant stress due to earth movements were ephemerally, but critically, increased during earth tremors. Terzaghi (1950) has emphasised the importance of the horizontal components of earthquake motion in connection with landslides and, in general terms, the sudden release of strain energy in the rock resulting from a high horizontal stress field could induce both failure of the surface rock and sufficient shaking to produce instability.

Such an explanation accounts for the varied results of A-tent development, involving mainly expansion but in places some contraction, and for the preferred location of angular tents and other forms described here on midslope where stresses are greatest on an arched (convex) surface (Price, 1966). Arched and angular forms are in these terms members of a continuum. Given that the near-surface granite is subdivided into layers (it is divided by sheeting joints and flaggy partings), lateral pressure may result first in arching (a in Fig. 11.30). Further compression would cause a crestal fracture to develop (b in Fig. 11.30), and still greater pressure the marginal fractures (c in Fig. 11.30). Thus the various arched and angular forms can be explained as stages in compression.

If the forms are related to a catastrophic event or events such as earthquakes or tremors, the wedges, both vertical and lateral, are explained in terms of a sudden, possibly short-term, increase in compressive stress. The resultant shaking would explain how friction and inertia have been overcome, as is implicit in the slipped slabs. There is evidence in the form of mineral changes due to friction (Twidale, 1964; Jaeger, (1965) of the granite exposed along the sheeting joints of differential movement along the planes, so that the lateral wedges can be understood as due to such dislocation.

It is notable that on Lightburn Rocks, where A-tents are well developed, there are also many recently developed fractures, including some vertical dislocations between juxtaposed slabs.

This assemblage of forms is best developed on northern Eyre Peninsula, in a zone bounded by faults (see Bourne *et al.*, 1974) and known to be seismically active (Sutton and White, 1968). The late D.J. Sutton (pers. comm.) reported

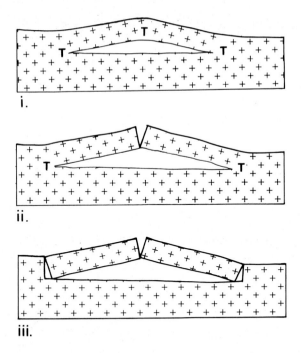

Fig. 11.30. Stages in the development of arched and angular tents. (T - tension)

that a small portable seismograph set up on Mt Wudinna for twelve hours in late 1975 recorded two seismic events in that short period. Moreover, Corrobinnie Hill, which may be interpreted as a horst block located within a complex fault zone (see Twidale, 1971a, p. 123; Bourne *et al*., 1974), is notable for the number of split and fallen blocks, possibly as a result of frequent earth tremors.

Thus a distinctive suite of minor landforms is developed on granitic rocks. It is best understood as associated with the release of compressive stress, and constitutes further evidence concerning the condition of the rock masses on which it is evolved.

PART IV

OVERVIEW

CHAPTER 12

DISCUSSION AND CONCLUSIONS

In many of the foregoing chapters reference has been made to forms that are developed on granitic materials and that are, or are allegedly, associated with particular climatic regimes. In this respect granite outcrops have been treated in the same manner as the rest of the land surface, for, from the earliest years of the scientific study of scenery, not only individual landforms but also entire assemblages or landscapes have been interpreted as reflecting the climatic conditions under which they evolved. The argument has commonly been extended to account for relic forms, or features that have supposedly been developed by processes and climates that no longer obtain in the areas to which reference is made. In many instances, of course, the argument is soundly based. The Permian glacial forms and deposits characteristic of many parts of Gondwanaland are clearly alien to the environments in which they are now found, and the landscapes of many parts of northern Europe and North America are incomprehensible without reference to the glaciations of the recent past.

Other allegedly inherited forms, including many of those developed in granitic rocks, are less securely based in reason. Pediments have been taken as evidence of former aridity or semiaridity and peneplains of humid tropical landscapes (e.g. Baulig, 1956). Some granite forms have been used as climatic indicators, boulders, for instance, being consistently, though in many instances mistakenly, taken for glacial erratics. Fluted surfaces, too, were associated with glaciation. Inselbergs and pediments, both individually and considered together in *Inselberglandschaften* were, and still are, widely treated as of tropical and of arid, or more popularly semiarid (savanna), provenance. Where they are found beyond the confines of such regions they are taken to be inherited (e.g. Galli-Olivier, 1967).

In this connection it may be of interest to recount that some years ago the present writer had the temerity to suggest, in the course of the discussion section of a paper submitted for publication in a well-known and prestigious professional journal, that the 'tors' of Dartmoor and other high plains of southwestern England ought, perhaps, be regarded as temperate inselbergs fringed by pediments, and with a piedmont angle or nick well developed and preserved. The distinguished referee, properly anonymous, though readily identified by his calligraphy and superb style, did not disagree, but, intent on seeing the paper in print and with its main thrust unclouded by incidental controversies, prudently suggested that the aside was provocative (as, indeed, it was intended to be) and that the paper could well stand without that particular brief paragraph. He was, of course, right; the advice was taken and the paper published, but the episode well illustrates the attitudes prevailing, then and now, as to the accepted environments of inselbergs and pediments.

Climatic zonation has long been a prominent, if not dominant, feature of geomorphological interpretation, and the analyses and explanations applied to granitic landscapes are no exceptions. Yet, even if only present conditions and distributions are taken into consideration, the patterns that emerge are far from simple.

Some forms, such as rock doughnuts and pedestals, though due to weathering and erosion, have not been reported in sufficient numbers for any patterns to be discernible, but many forms are known from enough areas, and in large enough numbers, to form significant populations. Several types of distribution can be identified. Some features, like rock basins and boulders, are so widely and well developed that they can be regarded as virtually ubiquitous. A few forms are essentially restricted to a particular climatic zone. Nubbins are best and most commonly developed in the humid tropics. Clitter, or frost riven plates and slabs, are restricted to cold lands. But such zonal forms are few in number.

Many other granitic forms are widely distributed, yet are seemingly favoured by some climates and processes. Thus bornhardts are especially well developed in savanna lands, but are also well represented in truly arid, humid tropical and subtropical, monsoon, temperate and cold areas. Again, pediments are well developed on semiarid regions, but have also been reported from temperate and cool lands.

The distribution of several granite forms can only be described as peculiar or odd, considered in terms of conventionally defined climatic regions. Castellated inselbergs, for example, are found in abundance in cold (both high latitude and high altitude) regions, but also occur in excellent shape in such warm, seasonally arid regions as southern Africa and central Australia. Tafoni have been reported from warm arid and semiarid lands, from Mediterranean regions, and from periglacial lands, as well as coastal zones that experience a hot, dry season; but they are not found well or widely developed in the humid tropics or in humid temperate regions. All-slopes topography is found in cold lands, in the humid subtropics, and in such arid uplands as the Sinai and the northern Flinders Ranges in South Australia. Fluted surfaces have been reported from glaciated regions, the humid tropics and arid and semiarid lands, as well as coastal areas, but are apparently absent from temperate regions. Peneplains have been described developed in weak (weathered) granite in a considerable variety of climatic contexts.

Finally, a few forms are, seemingly, restricted to particular, specific areas. Thus flared slopes, though noted in many and varied locations, attain their best and most prolific development in southern Australia. Again, some of the forms claimed to be related to the release of stress (A-tents, displaced slabs and wedges) have so far been reported primarily from southern Australia and especial-ly northwestern Eyre Peninsula; but they do occur elsewhere and this apparently restricted distribution may reflect more a lack of observation and report than any significant genetic factor.

Climatic interpretations of granite forms are complicated by a number of factors that render the process unreal and frankly simplistic. To explain the present distribution of bornhardts, for example, in terms of the savanna morphogenetic system favoured by many workers (of whom Bloom, 1978, p.324, may be cited as a recent example) calls for extensions of semiaridity far beyond present limits and far beyond the present data. There is no evidence of savanna conditions having obtained in southeastern Brazil, for instance, during the later Cainozoic which is the relevant period of bornhardt genesis in this locality. There is no suggestion of savanna conditions in the recent past in northern Norway. Moreover, some of the supposed links between climate, process and landforms are uncertain. It is far from certain that bornhardts have formed only in savanna conditions. The *morros* of southeastern Brazil have evolved mainly under the influence of warm humid climates, and the forms of Mesozoic and early Cainozoic age described from various parts of southern and central Australia are associated with various types of duricrusted plains that probably developed in warm humid or subhumid climates.

Distributions can be deceptive. Flared slopes, for instance, are best preserved in southern Australia, and best of all in alkaline environments, the inselbergs on the lower slopes of which they are sculptured being typically surrounded by plains capped by calcrete or by saline soils. It is tempting to attribute pronounced scarp foot weathering to which flared slopes owe their origin to strong preferential attack of the acidic granites by alkaline waters, much in the fashion implied by Joly (1901) and Mason (1966, p. 142). Such marked attack by alkaline groundwaters undoubtedly takes place, but there are several objections to the suggestion in the context of granitic environments. Granite weathering takes place in conditions of low, or moderate, pH. Again, the flared slopes of southern Australia were initiated beneath land surfaces of various ages (Mesozoic, earlier Cainozoic), but consistently under warm humid or subhumid conditions not conducive to the formation of alkaline soils. Then, if high soil pH were the essential factor, flared slopes ought to be found in all the arid and semiarid lands the world over; and they have not been so reported.

Other factors have to be considered. Not only are massive rocks essential to the development of the forms but extended periods of landscape stability are necessary for the concentrated scarp foot weathering to take place (Twidale, 1972b). The shield areas of southern Australia have been relatively stable, subject only to epeirogenic movements for much of Phanerozoic time. On the other hand, these vertical movements have induced phases of stream rejuvenation and landscape revival, so that the concave weathering fronts have been exposed, and the aridity of the later Cainozoic has contributed to the preservation of the elegant forms. In these terms, of course, flared slopes ought to be well developed in several parts of southern and central Africa, and the Brazilian Shield; and they may be, either unreported or not yet exposed, awaiting rejuvenation to attain the deep interiors of the large land masses of which they are part.

Another complicating factor, clearly appreciated by Wilhelmy (1958, pp. 205-210), is convergence or equifinality, that is, the production of similar forms by different processes. Boulders, are of the most common and characteristic of all granite forms, may be taken as an example. Most have evolved through differential fracture-controlled subsurface weathering. This weathering may have taken place over a period of some millions of years, during which time climatic changes have almost certainly occurred. But whatever reactions dominated the weathering process at various times, whether chemical or mechanical, weathering prevailed, whether solution, hydration or hydrolysis was dominant, whatever clays were produced at various times, the overall effect was to produce corestones set in grus or another form of weathered matrix. Thus it can, on general grounds, be argued that various climatically controlled weathering processes have, together, produced a particular form or assemblage. It is only rarely possible to relate particular causes and effects, but in the instance cited all processes worked toward the same end, so that the form, in this instance the boulder, can be regarded as a convergent form, equivalent to the homeomorph of the palaeontologist.

Mushroom rocks such as those illustrated in Figs. 9.12 (b), (c) and (e) provide another example. All are due to the attack, by soil moisture, of the unexposed parts of large corestones. The exact process involved probably varied from one area to another and even from site to site, as well as from time to time, and the processes responsible for the exposure of the residual masses shaped by weathering, has certainly varied, from the wind in one example to frost action in another, though with important effects due to gravity in all of the examples under discussion - the simple falling away of loose particles no longer buttressed by an adjacent mass of regolithic debris.

Bornhardts, too, have evidently evolved in various ways, and in a wide range of climatic environments. In addition, however, structural convergence has also occurred, for various controls - lithology, fracture density, perhaps tectonism - have acted to produce similar effects, namely protuberances which have in time become upstanding hills. Although the causation varies from site to site, the end products are morphologically similar.

Such convergent forms differ from polygenetic features in that the latter bear the imprint of clearly contrasted processes, as, for instance, in northwestern Europe where fluvial forms are superimposed on glacial features, with even older features and forms such as palaeosurfaces and deep weathering profiles still evident in places, and with structural controls significant, as always.

Climatic inheritance and convergence, then, are two factors that must be considered when the distribution of granite forms is examined. The susceptibility of granite to moisture attack, combined with the virtual ubiquity of groundwaters, is another that has imposed uniformity and azonality. The physical and chemical characteristics of granite render it particularly susceptible to moisture attack but there is a marked contrast in permeability and hence in vulnerability as between fresh and weathered granite. The character of granitic rocks promotes the development of a sharp boundary, a clear weathering front, between weathered and fresh rock. There is a clearly delineated interface between granite affected by moisture and granite that has not been penetrated.

Such subsurface moisture attack has exploited various weaknesses in the bedrock with which it has come into contact, bringing into relief the more resistant compartments or masses and lowering the weathering front in the vicinity of zones and planes that are vulnerable to attack. The presence of runnels in the weathering front on the flanks of bedrock protuberances argues a funnelling of water from high to low points in the covered relief and suggests that positive feedback mechanisms operate in the subsurface as well as on exposed forms.

Over wide areas, then, granite has been weathered so that the fresh rock has come to be masked by a more-or-less thin veneer of grus, clay and transported, introduced material that, together, form the regolith. Once rivers and streams begin to incise their beds, either in response to baselevel lowering or to reduced calibre of bedrock resulting from weathering, two types of situation can be envisaged (see Ollier, 1960). On the one hand, the streams have cut down but remain within the regolith: here, plains underlain by and eroded in previously weathered rock have been formed, interrupted only by upstanding projections of the weathering front that have been exposed as etch features. The plains vary in detail according to proximity to uplands and the duration and perfection of planation but such plains are widely developed, though in practice it is difficult to distinguish between weathering that preceded planation and that which occurred at the same time as planation, or followed the formation of, the plains.

Where, on the other hand, stream incision has extended to or below the level of the weathering front, the regolith has been stripped away completely to expose the weathering front as an etch plain complete with minor depressions and channels.

Subsurface weathering has proceeded almost regardless of climatic conditions at the surface. Only in extreme nival conditions may subsurface weathering have been interrupted. Thus, over wide areas a covered and differentiated landscape has been formed beneath the land surface, a landscape the morphology of which transgresses climatic boundaries, and one which over wide areas has been wholly or partially exposed to form part of the present scenery. Forms such as bornhardts and other inselbergs, boulders, runnels and depressions, flared slopes and scarp foot depressions, tafoni and pitting, as well as platforms and extensive plains are, as has been documented and argued in earlier chapters, all etch forms. The development of the weathering front and the nature of the reactions that take place there are, therefore, fundamental to an understanding of many common granite features.

The regolith and weathering front due to subsurface weathering have, of course, taken time to develop. Just as the immensity of time is crucial to any acceptance of the possibility of organic evolution (Eiseley (1961, p. 65) has said of James Hutton that he 'discovered ... time', and that this was an essential precursor to Charles Darwin's famous theory), so is an appreciation of the antiquity of many granite forms essential to their explanation. The granitic shield lands are the most ancient nuclei of the continents, and plutons are a significant component of many orogens, ancient and relatively recent. Both have been subjected to immense crustal stresses and the older granite masses, especially, have been subjected to long-continued moisture attack as they have emerged into the groundwater zone following uplift and erosion. But even comparatively youthful granite masses have been weathered, reflecting the vulnerability of granitic rocks to moisture attack.

Additionally, for long periods of the Mesozoic and the early Cainozoic (all but the last two millions of years) warm humid or subhumid conditions prevailed over wide areas of the continents, even in quite high latitudes, so that conditions were particularly conducive to deep intense weathering, especially in stable areas like the shields. This combination of ample time and the vulnerability of granite to moisture attack, plus the ubiquity of groundwaters now and in the past, explains why plains are the typical granite form. In areas of particular stability almost all structural variations have been subdued, resulting in such extraordinary features as the ultiplains of the interiors of Western Australia and southwestern Africa. But, even here, some resistant compartments persist (Fig. 1.14), indicating real and significant physical or chemical attributes that have permitted persistence and survival against all odds over prolonged periods, and suggesting that the reasons some compartments are resistant ought to be investigated in detail. In particular, stress conditions and patterns, and associated landforms would repay further study. Their relations with plate tectonics in particular await analysis and contemplation.

The etch forms developed in granite have two ages, one related to the period when the masses were shaped by subsurface moisture attack, the other to the period of exposure. The two may be similar (for weathering and erosion may proceed simultaneously or virtually so) or they may be widely different. Thus, to take an example of the latter, if Worden and Compston (1973) are correct in identifying the weathering of the Martindale granite of central Western Australia as a Precambrian, indeed, as an Archaean, event, the first stage in the development of the low, large-radius domes took place some 2000 Ma before the second stage involving the late Cainozoic (post-lateritic) exposure of the weathering front. Again, Barbier (1967) has described from the Tassili Mountains of southern Algeria a landscape of inselbergs formed by differential fracture-controlled weathering that occurred in later Precambrian or earliest Cambrian times, i.e., before the deposition of then Tassili Sandstone, though the exploitation of the differentially weathered granite basement has not taken place until relatively recently. In many parts of Australia the deep weathering that resulted in the formation of laterite in southern South Australia and in the differential subsurface wathering of some granitic masses occurred in the early Mesozoic, and that associated with silcrete development in the early Cainozoic. But whatever the age or age-range of the weathering, it followed a common pattern and similar forms resulted.

The second stage of development, the exposure of the weathering front, has also taken place at various times and under varied conditions. Although features are still in process of exposure, others have been uncovered for astoundingly long periods of geological time. Early Cainozoic forms are fairly common, and features of Mesozoic age have been suggested. Their survival is yet another potent field of investigation (see Twidale, 1976a). The agencies responsible for the uncovering of the weathering front vary from place to place and no doubt time to time, and they impose their own characteristics in detail; but whether rivers or waves or glaciers are responsible, the major features etched in the granite bedrocks are basically similar.

Finally, azonality and a measure of commonality is imposed by the structural factors that have already been mentioned in several connections and especially with regard to boulders and bornhardts. But they are of prime significance throughout the granitic domain. Fresh granite is of low porosity and permeability. Water infiltrates into the rock mass only very slowly, the main means of access being provided by fractures, by faults and joints, which thereby assume a crucial significance in the pattern of weathering and, hence, of landform development. As Rognan (1967, p. 275) so succinctly put it, 'Les diaclases gardent un rôle primordial en influençant le modelé granitique'. In addition, however, micropartings such as crystal cleavages, boundaries and dislocations are not unimportant and may well *in toto* be as effective as the more readily detectable major fractures with which, however, they may well be distribution-ally coincident and genetically related. Water that infiltrates along these partings causes the rock adjacent to the fractures to be altered, so creating a marked contrast between the friable grus and the fresh, cohesive rock. Once weathered, the granite is much more permeable, so that the rate of weathering accelerates: another positive feedback or reinforcement mechanism. The contrast between fresh and weathered rock is exploited by erosional agencies, so that the patterns of fractures that are related to crustal stress are closely reflected in patterns of landform and topography, regardless of climatic conditions.

All this is not to imply that climatically induced morphogenetic systems have had no influence on the shaping of granite forms or that none has climatic significance. Glaciated and nival regions stand out from others, regardless of the lithological base, because of their climatically imposed peculiarities. And they are indeed distinctive, with their tendency to angularity, the common development of spalling, the formation of clitter and talus slopes. Such sculpted features as cirques, horns, U-shaped valleys, as well as various glacial and fluvioglacial depositional forms are characteristic *per se*, and are deve-loped in a wide variety of lithological environments including granites (Figs. 5.4, 6.2 and 6.6). Yet the landforms of cold lands are exceptional; no other suite of erosional forms is as readily distinguished. Elsewhere, the distribu-tions are less distinct and more confused. Indeed, the angular all-slopes, so well and widely evolved in cold lands, are emulated in both the humid subtropics (e.g. southeastern Brazil) and in truly arid low latitude regions in the Sinai and in the northern Flinders. Moreover, rounded domical forms said to be characteristic of savanna regions are found also in cold regions in Norway, the Yosemite and even Greenland.

The humid warm assemblage of plains and nubbins is also distinctive, and Oberlander's (1972) suggestion that the nubbins or knolls of the Mojave Desert are relic from an earlier (early Cainozoic) phase of wetter climate is well founded. But it is also opportune to point out that the tropical humid system may well have prevailed over wide areas for some scores of millions of years. The landscape may have evolved under conditions that were unchanged for long periods. The landscapes of the humid tropics, survivors of the general or widespread climatic state of the later Mesozoic and much of the earlier Cainozoic; they may represent the only morphogenetically pure system.

Thus climate is important in places but it is only one of several controls that have influenced the development of landforms on granitic outcrops. Its effects are generally overshadowed by the tendencies to common development imposed by structural factors, by the fact that many forms have developed at some stage under warm humid conditions that were widespread through later Phanerozoic time, and by the fact that many granite forms are of etch character and evolved by subsurface weathering attack under conditions that were similar over wide areas. The evolution of granite landforms is complex rather than simple, but as Lahee (1909) pointed out, complexity is not synonymous with chaos any more than uniformity and simplicity go hand in hand. Though there is a healthy disagreement about the origin of many granite forms, many are susceptible of rational analysis. Granite landscapes, like all others, are a function of Davis' (1909) famous triad of structure, process and time, but past processes, largely involving groundwaters, and the shaping of forms large and small at the weathering front, here assume a greater significance than elsewhere.

REFERENCES CITED

Ackermann, E., 1962. Büssersteine - Zeugen Vorzeitlicher Grundwasserschwankungen, Z. Geomorph. 6: 148-182.

Ackermann, H.D., 1974. Shallow seismic compressional and shear wave refraction and electrical resistivity investigations at Rocky Flats, Jefferson County, Colorado. U.S. geol. Surv. J. Res. 2: 421-430.

Agassiz, L., 1865. On the Drift in Brazil, and on decomposed rocks under the Drift. Amer. J. Sci. & Arts 40: 389-390.

Ahnert, F., 1967. The role of the equilibrium concept in the interpretation of landforms of fluvial erosion and deposition. Proc. Symp. l'évolution des versants (Liège) 40: 23-41.

Akagi, Y., 1972. Pediment morphology in Japan. Fukuoka Univ. Educ. Bull. 21: 1-63.

Akagi, Y., 1974. Pediment on the Taean Peninsula and the Yeogsan River Basins, Korea. Sci. Rept. Tohoku Univ. 7th. Ser. (Geogr.) 24: 183-197.

Alexander, F.E.S., 1959. Observations on tropical weathering: a study of the movement of iron, aluminium and silicon in weathering rocks at Singapore. Quart. J. geol. Soc. London, 115: 123-142.

Allen, J.R.L., 1970. Physical Processes of Sedimentation. George Allen & Unwin, London.

Amaral, I.do, 1973. Formas de 'Inselberge' (ou montes-ilhas) e de meteorização superficial e profunda em rochas graniticas do Deserto de Moçãmedas (Angola) ne margem direita do rio Curoca. Garcia de Orta, Ser. Géogr. 1: 1-34.

Ambrose, J.W., 1964. Exhumed paleoplains of the Precambrian Shield of North America. Amer. J. Sci. 262: 817-857.

Anderson, A.L., 1931. Geology and mineral resources of eastern Cassia County, Idaho. Idaho Bur. Mines & Geol. Bull. 14.

Andrée, K., 1912. Die geologische Bedeutung des Wachstumdrucks Kristallisierender Substanzen. Geol. Rdsch. 3: 7-15.

Andrée, D.V. d', Fischer, R.L. and Fogelson, D.E., 1965. Prediction of compressive strength from other rock properties. U.S. Dept. Interior, Bureau of Mines, Rept. of Investig. 6702.

Anhaeusser, C.R., 1973. The geology and geochemistry of the Archaean granites and gneisses of the Johannesburg-Pretoria Dome. In: Symposium on granites, gneisses and related rocks, L.A. (Lister (Editor). Geol. Soc. S. Af. Sp. Pub. 3: 361-385.

Arnold, J., 1980. Death in the Rockies. NSS News (September), pp. 209-215.

Badgley, P., 1965. Structural and Tectonic Principles. Harper & Row, New York, Weatherhill, Tokyo.

Bagnold, R.A., 1933. A further journey through the Libyan Desert. Geogrl. J. 82: 211-235.

Bain, A.D.N., 1923. The formation of inselberge. Geol. Mag. 60: 97-107.

Bakker, J.P., 1958. Zur granitverwitterung und Methodik der Inselbergforschung in Surinam. Deutsch. Geographentag Wurzburg, 29: 121-131.

Bakker, J.P., Kwaad, F.J.P.M. and Müller, H.J., 1968. Einige vorlaufige Bermerkungen über Salz - und Tonsprengung besonders in Hinblick auf Granit. Przegląd Geograficzny 40: 387-399.

Balk, R., 1937. Structural behaviour of igneous rocks. Geol. Soc. Amer. Mem. 5.

Barbeau, J. and Gèze, B., 1957. Les coupoles granitiques et rhyolitiques de la région de Fort-Lamy (Tschad). C.R. Somm. et Bull. Soc. géol. France. (Ser. 6), 7: 341-351.

Barbier, R., 1957. Aménagements hydroélectriques dans le sud du Brésil. C.R. Somm. et Bull. Soc. géol. France 6: 877-892.

Barbier, R., 1967. Nouvelles réflexions sur le problème des 'pains de sucre' à pupos d'observations dans le Tassile N'Ajjer (Algerie). Trav. Lab. Géol. Fac. Sc. Grenoble 43: 15-21

Bardin, V.I., 1963. Zonality of periglacial phenomena in the mountains of Queen Maud Land. *Sov. Anarct. Exped. Inform. Bull.* 5: 4-6.

Barrère, P., 1952. Le relief des massifs granitiques du Néouvielle, de Cauterets et de Panticosa. *Rev. Géogr. Pyrénées et Sud-Ouest* 23: 69-98.

Barrère, P., 1968. Le relief des Pyrénées centrales occidentales. *J. d'Etudes Pau-Biarritz.* 194: 31-52.

Barton, D.C., 1916. Notes on the disintegration of granite in Egypt. *J. Geol.* 24: 382-393.

Bartrum, J.A., 1936. Honeycomb weathering of rocks near the shoreline. *N.Z.J. Sci. Tech.* 18: 593-600.

Bateman, P.C., *et al.,* 1963. The Sierra Nevada Batholith. A synthesis of recent work across the central part. *U.S. geol. Surv. Prof. Paper* 414-D.

Bateman, P.C. and Eaton, J.P.,1967. Sierra Nevada Batholith. *Science* 158: 1407-1417.

Bateman, P.C. and Wahrhaftig, C., 1966. Geology of the Sierra Nevada. In: E.H. Bailey (Editor) *U.S. geol. Surv. Bull.* 190.

Bauer, M., 1898. Beitrage zur Geologie der Seychellen. *Neues Jahrb. f. Min. Geol. Pal.* (B) 64: 79-146.

Baulig, H., 1952. Surfaces d'aplanissement. *Ann. Géogr.* 61: 161-183, 245-262.

Baulig, H., 1956. Pénéplaines et pédiplaines. *Bull. Soc. Belge d'Études géogr.* 25: 25-58.

Beche, H.T. de la, 1839. Report on the geology of Cornwall, Devon and West Somerset. *Geol. Surv. Engl. Wales.*

Beche, H.T. de la, 1853. *The Geological Observer.* (2nd edition) Longmans, Brown, Green and Longmans, London.

Becker, G.F. and Day, A.L., 1905. The linear force of growing crystals. *Proc. Wash. Acad. Sci.* 7: 283-288.

Becker, G.F. and Day, A.L., 1916. Notes on the linear force of growing crystals. *J. Geol.* 24: 313-333.

Bendefy, L., 1959. Niveauanderungen in Raum von Transdanubian auf Grund Zeitgemasser Fein ein wagungen. *Acta tech. Hung.* 23: 167-168.

Berger, A.R. and Pitcher, W.S., 1970. Structures in granitic rocks: a commentary and a critique on granite tectonics. *Proc. Geol. Ass. (London)* 81: 441-461.

Berliat, K., 1965. Exploratory drilling for water in the sandplain country 40 miles east of Hyden. *Geological Survey of W. Aust. Ann. Rept. (1964),* pp. 8-9.

Birot, P., 1950. Notes sur le problemè de la désagrégation des roches crystal-lenes. *Rev. Géomorph. Dynam.* 1: 271-276.

Birot, P., 1952. Le relief granitique dans le nord-ouest de la Péninsule Ibérique. *Proc. 17th Int. Cong. I.G.U.,* Washington, pp. 301-303.

Birot, P., 1954. Désagrégation des roches cristallines sons l'action des sels. *C.R. Acad. Sci.* 238: 1145-1146.

Birot, P., 1958. Les dômes crystallins. *C.N.R.S. Mem. & Doc.,* pp. 7-34.

Birot, P., Godard, A., Petit, M. and Ters, M., 1974. Contribution à l'étude des surfaces d'aplanissement et de l'érosion differentielle dans le Transvaal septentrional et oriental (Afrique du Sud). *Rev. Géogr. Phys. Géol. Dynam.* 16: 421-554.

Birot, P. and Joly, F., 1952. Observations sur les glacis d'érosion et les reliefs granitiques au Maroc. *Mem. Doc. Cent. Docum. cartogr. géogr.* 6: 35-57.

Blackwelder, E., 1925. Exfoliation as a phase of rock weathering. *J. Geol.* 33: 793-806.

Blackwelder, E., 1927. Fire as an agent in rock weathering. *J. Geol.* 35: 134-140.

Blackwelder, E., 1929. Cavernous rock surfaces of the desert. *Amer. J. Sci.* 17: 393-399.

Blackwelder, E., 1931. Desert plains. *J. Geol.* 39: 133-140.

Blackwelder, E., 1933. The insolation hypothesis of rock weathering. *Amer. J. Sci.* 26: 97-113.

Blackwelder, E., 1948. Historical significance of desert lacquer. *Geol. Soc. Amer. Bull.* 59: 1367.

Blank, H.R., 1951a. Exfoliation and granite weathering on granite domes in central Texas. *Texas J. Sci.* 3: 376-390.

Blank, H.R., 1951b. "Rock doughnuts", a product of granite weathering. *Amer. J. Sci.* 249: 822-829.

Bloom, A.L., 1978. *Geomorphology. A Systematic Analysis of Late Cenozoic Landforms.* Prentice-Hall, New Jersey.

Blyth, F.G.H., 1957. The Lustleigh fault in northeast Dartmoor. *Geol. Mag.* 94: 291-296.

Blyth, F.G.H., 1962. The structure of the northeastern tract of the Dartmoor granite. *Quart. J. geol. Soc. London* 118: 435-453.

Boase, W., 1834. *Treatise on Primary Geology.* Longman, Rees, Orme, Brown, Green and Longman, London.

Bocquier, G., Rognon, P., Paquet, H. and Millot, G., 1977. Géochimie de la surface et formes du relief. II. Interprétation pédologiques des dépressions annulaires entourant certains inselbergs. *Sci. Géol. Bull.* 30: 245-253.

Boland, J.N., McLaren, A.C. and Hobbs, B.E. 1971. Dislocations associated with optical features on naturally-deformed olivine. *Contr. Miner. Petrol.* 30: 53-63.

Booth, B., 1968. Petrogenic significance of alkali feldspar megacrysts and their inclusions in Cornubian granites. *Nature (London)* 217 (5133): 1036-1038.

Borlase, W., 1754. *Observations on the Antiquities, Historical and Monumental, of the County of Cornwall.* Bowyer and Nichols, London.

Bornhardt, W., 1900. *Zur Oberflächengestaltung und Geologie Deutsch Ostafrikas.* Reimer, Berlin.

Bott, M.H.P., 1953. Negative gravity anomalies over 'acid intrusions' and their relation to the structure of the earth's crust. *Geol. Mag.* 90: 257-267.

Bott, M.H.P., 1956. A geophysical study of the granite problem. *Quart. J. geol. Soc. London* 412: 45-62.

Bott, M.H.P. and Smithson, S.B., 1967. Gravity investigations of subsurface. Shape and mass distributions of granite batholiths. *Geol. Soc. Amer. Bull.* 78: 859-878.

Bourcart, J., 1957. *L'Érosion des Continents.* Colin, Paris.

Bourne, J.A., Twidale, C.R. and Smith, D.M., 1974. The Corrobinnie Depression, Eyre Peninsula, South Australia. *Trans. Royal. Soc. S. Aust.* 98: 139-152.

Boyé, M., 1950. *Glaciaire et Périglaciaire de l'Ata Sund, Nord-Oriental Groenland.* Hermann, Paris.

Boyé, M. and Fritsch, P., 1973. Dégagement artificiel d'un dôme crystallin au Sud-Cameroun. *Trav. & Doc. Géogr. Trop.* 8: 69-94.

Bradley, W.C., 1963. Large-scale exfoliation in massive sandstones of the Colorado Plateau. *Geol. Soc. Amer. Bull.* 74: 519-528.

Bradley, W.C., Hutton, J.T. and Twidale, C.R., 1978. Role of salts in development of granitic tafoni, South Australia. *J. Geol.* 86: 647-654.

Bradley, W.C., Hutton, J.T. and Twidale, C.R., 1979. Role of salts in development of granitic tafoni, South Australia: a reply. *J. Geol.* 87: 121-122.

Brajnikov, B., 1953. Les pains-de-sucre du Brésil: sont-ils enracinés? *C.r. somm. & Bull. Soc. Géol. France* 6: 267-269.

Brammall, A., 1926. The Dartmoor granite. *Proc. Geol. Ass. (London)* 37: 251-277.

Brammall, A. and Harwood, H.F., 1932. The Dartmoor granites: their genetic relationships. *Quart. J. geol. Soc. London* 88: 171-237.

Branner, J.C., 1896. Decomposition of rocks in Brazil. *Geol. Soc. Amer. Bull.* 7: 255-314.

Branner, J.C., 1913. The fluting and pitting of granites in the tropics. *Proc. Amer. Phil. Soc.* 52: 163-174.

Bremer, H., 1965. Ayers Rock, ein Beispiel für klimagenetische Morphologie. *Z. Geomorph.* 9: 249-284.

Bremer, H. and Jennings, J.N. (Editors), 1978. Inselbergs/Inselberge. *Z. Geomorph. Suppl-band* 31.

Bridgman, P.W., 1912. Water, in the liquid and five solid forms, under pressure. *Proc. Amer. Acad. Arts & Sci.* 47: 441-558.

Bridgman, P.W., 1938. Reflections on rupture. *J. appl. Phys.* 9: 517-528.

Brown, D.A., Campbell, K.S.W. and Crook, K.A.W., 1968. *The Geological Evolution of Australia and New Zealand.* Pergamon, London.

Browne, W.R., 1964. Grey billy and the age of tor topography in Monaro, N.S.W. *Proc. Linn. Soc. N.S.W.* 89: 322-325.

Bruckner, W.D., 1966. Salt weathering and inselbergs, *Nature* 210 (5038), p. 832.

Brunner, F.K. and Scheidegger, A.E., 1973. Exfoliation. *Rock Mechanics* 5: 43-62.

Brunsden, D., 1964. The origin of decomposed granite on Dartmoor, pp.97-116. In: I.G. Simmons (Editor) *Dartmoor Essays.* Dev. Assoc. Adv. Sci. Lit. & Art, Torquay.

Bryan, K., 1922. Erosion and sedimentation in the Pagago Country, Arizona. *U.S. Geol. Surv. Bull.* 730: 19-90.

Buckley, H.E., 1951. *Crystal Growth.* Wiley, New York.

Büdel, J., 1957. Die 'Doppelten Einebnungsflächen' in den feuchten Tropen. *Z. Geomorph.* 1: 201-228.

Büdel, J., 1959. Morphogenese des Festlandes in Abhangigkeit von den Klimazonen. *Die Naturwissenschaften* 48: 313-318.

Büdel, J., 1963. Klima-genetische Geomorphologie. *Geogr. Rdsch.* 7: 269-286.

Büdel, J., 1977. *Klima-Geomorphologie.* Borntraeger, Berlin.

Bur. Recherches Géol. Minières, 1968. *Carte Géologique de la France 1:1 000 000 (Feuille Nord)* Orleans.

Bulow, K., 1942. Karrenbildung in kristallen Gesteinen? *Z. Deutsch. Geol. Geselt.* 94: 44-46.

Cailleux, A., 1953. Tafonis et érosion alvéolaire. *Cah. géol. Thoiry* 16-17: 130-133.

Cailleux, A., 1962. Études de géologie au détroit de McMurdo (Antarctique). *Comm. Natl. Français Rech. Antarctiques* 1.

Caine, N., 1967. The tors of Ben Lomond, Tasmania. *Z. Geomorph.* 11: 418-429.

Caldcleugh, A., 1829. On the geology of Rio de Janeiro. *Trans. geol. Soc. London* 2: 69-72.

Calkin, P. and Cailleux, A., 1962. A quantitative study of cavernous weathering (tafonis) and its application to glacial chronology in Victoria Valley, Antarctica. *Z. Geomorph.* 6: 317-324.

Callen, R.A. and Tedford, R.H., 1976. New Late Cainozoic rock units and depositional environments, Lake Frome area, South Australia. *Trans. Royal Soc. S. Aust.* 100: 125-167.

Cameron, J., 1945. Structural features of the grey granites of Aberdeenshire. *Geol. Mag.* 82: 189-204.

Carey, S.W., 1955. The orocline concept in geotectonics. I. *Papers Proc. Royal Soc. Tas.* 89: 255-288.

Carlé, W., 1941. Karrenbildung im Granit der galischen Kuste bei Vigo (Norwestspanien) *Geol. Meere Binnengewasser* 5: 55-63.

Carl, J.D. and Amstutz, G.C., 1958. Three-dimensional Liesegang rings by diffusion in a colloidal matrix, and their significance for the interpretation of geological phenomena. *Geol. Soc. Amer. Bull.* 69: 1467-1468.

Carnegie, D.W., 1898. *Spinifex and sand: a narrative of five years' pioneering and exploration in Western Australia.* Pearson, London.

Chapman, C.A., 1956. The control of jointing by topography. *J. Geol.* 66: 552-558.

Chapman, C.A. and Greenfield, M.A., 1949. Spheroidal weathering of igneous rocks. *Amer. J. Sci.* 247: 407-429.

Chayes, F., 1957. A provisional reclassification of granite. *Geol. Mag.* 94: 58-68.

Choffat, P., 1895-96. Notes sur l'érosion en Portugal. I. Sur quelques cas d'érosion atmosphérique dans les granites du Minho (Tafoni). *Trabalhos Geol. Portugal., Commun.* 3: 17-22.

Choubert, B., 1949. *Géologie et Petrographie de la Guyane Française.* O.R.S.O-M., Paris.

Choubert, B., 1974. Le Précambrien des Guyanes. *Mem. B.R.G.M.* 81.

Clayton, R.W., 1956. Linear depressions *(Bergfussniederungen)* in savannah land-scapes. *Geogrl. Stud.* 3: 102-126.

Cloos, E., 1936. Der Sierra-Nevada pluton in Californien. *N. Jahrb. f. Min. Geol. Pal.* 76B: 355-450.

Cloos, E., 1955. Experimental analysis of fracture patterns. *Geol. Soc. Amer. Bull.* 66: 241-256.

Cloos, H., 1922. *Der Gebirgsbau Schlesiens und die Stellung seiner Bodenschätze.* Borntraeger, Berlin.

Cloos, H., 1923. Das Batholithenproblem. *Fortschr. d. Geol. u. Pal.* 1: 1-80.
Cloos, H., 1929. Zur Mechanik der Randzonen von Gletscher, Schollen und Plutonen, *Geol. Rdsch.* 20: 66-75.
Cloos, H., 1936. Plutone und Ihre Stellung im Rahmen der Krustenbewegungen. *16th Int. Geol. Congr. Rept.* 1: 235-253.
Cloudesley-Thompson, J.L., 1965. *Desert Life*. Pergamon, London.
Coates, D.F., 1964. Some cases of residual stress effects in engineering work, pp. 679-688. In: W.R. Judd (Editor) *The State of Stress in the Earth's Crust*. Elsevier, New York.
Cobbing, E.J. and Pitcher, W.S., 1972. The coastal batholith of central Peru. *Quart. J. geol. Soc. London* 128: 421-460.
Cole, W.F., 1959. Some aspects of the weathering of terracotta roofing tiles. *Austr. J. appl. Sci.* 10: 346-363.
Cooke, R.U., 1981. Salt weathering in deserts. *Proc. geol. Assoc. (London)*, 92: 1-16.
Cooke, R.U. and Smalley, I.J., 1968. Salt weathering in deserts. *Nature (London)* 200 (5173): 1226-1227.
Corbel, J., 1959. Vitesse d'érosion. *Z. Geomorph.* 3: 1-28.
Costin, A.B., 1950. Mass movements of the soil surface with special reference to the Monaro region (Pt. 2). *J. Soil Conserv. Serv. N.S.W.* 6: 73-80.
Cotton, C.A., 1941. *Landscape*. Whitcombe and Tombs, Christchurch.
Cotton, C.A., 1942. *Climatic Accidents*. Whitcombe and Tombs, Christchurch.
Cotton, C.A., 1948. *Landscape*. Whitcomb and Tomb, Christchurch.
Cowie, J.W., 1961. Contributions to the geology of North Greenland. *Med. Grønland*, 164.
Crickmay, C.H., 1933. The later stages of the cycle of erosion. *Geol. Mag.* 70: 337-347.
Crickmay, G.W., 1935. Granite pedestal rocks in the southern Appalachian piedmont. *J. Geol.* 43: 745-758.
Crowell, J.C., 1974. Sedimentation along the San Andreas Fault, California, pp.292-303. In: R.H. Dott and R.H. Shaver (Editors) *Modern and Ancient Geosynclinal Sedimentation, Soc. Economic Palsts. & Minsts. Spec. Publ.* 19.
Crowell, J.C., 1976. Implications of crustal stretching and shortening of coastal Ventura Basin, California. *Amer. Assoc. Petrol Geologists, Pac. Sect. Misc. Publ.* 24: 365-382.
Cunningham, F.F., 1969. The Crow Tors, Laramie Mountains, Wyoming, U.S.A. *Z. Geomorph.* 13: 56-74.
Czudek, T., Demek, J., Marvan, P., Panoš, V. and Raušer, J., 1964. Verwitterungs- und Abtragungsformen der Granits in der Böhmischen Masse. *Pet. geogr. Mitt.* 108: 182-192.
Dahl, E., 1961. Refugieproblemet og de kvartaer-geologiske metodene. *Svensk Naturkunskap* (Stockholm).
Dahl, E., 1963. Bomerkinger om refugieproblemet og de kvataegeologiske metodene. *Norsk. geol. Tidssk.* 43: 260-266.
Dahl, R., 1965. Plastically sculptured detail forms on rock surfaces in northern Nordland, Norway. *Geogr. Ann.* A47A: 83-140.
Dahl, R., 1966. Block fields, weathering pits and tor-like forms in the Narvik Mountains, Nordland, Norway. *Geogr. Ann.* A48A: 55-85.
Dahlke, J., 1970. Beobachtungen zum Phänomen der Hangversteilungen in Südwestaustralian. *Erdkunde* 24: 285-290.
Daily, B., Twidale, C.R. and Milnes, A.R., 1974. The age of the lateritised summit surface on Kangaroo Island and adjacent areas of South Australia. *J. geol. Soc. Aust.* 21: 387-392.
Dale, T.N., 1923. The commercial granites of New England. *U.S. geol. Surv. Bull.* 738.
Daly, R.A., 1912. Geology of the North American Cordillera at the forty-ninth parallel. *Geol. Surv. Canada, Mem.* 38.
Darwin, C.R., 1846. *Geological Observations on South America*. Smith, Elder, London.
Davis, S.N., 1963. Silica in streams and groundwater. *Amer. J. Sci.* 262: 870-891.

Davis, W.M., 1909. *Geographical Essays*. Dover, Boston.

Davis, W.M., 1938. Sheetfloods and streamfloods. *Geol. Soc. Amer. Bull.* 49: 1337-1416.

Dearman, W.R., 1963. Wrench-faulting in Cornwall and South Devon. *Proc. Geol. Ass. (London)* 74: 265-287.

Dearman, W.R., 1964. Dartmoor: its geological setting, pp. 1-29. In: I.G. Simmons (Editor) *Dartmoor Essays,* Dev. Assoc. Adv. Sci. Lit. & Art, Torquay.

Debon, F., 1972. *Massif Granitiques de Cauterets et Panticosa (Pyrénées Occidentales).* Expl. Notes. 1:50 000 Geol. Map. Bur. Recherches Geol. Min.

Demek, J., 1964a. Slope development in granite areas of Bohemian Massif (Czechoslovakia). *Z. Geomorph. Suppl.-Band.* 5: 82-106.

Demek, J., 1964b. Castle koppies and tors in the Bohemian Highland (Czechoslovakia). *Biul. Peryglac.* 14: 195-216.

Denaeyer, M.E., 1956. Actualité de l'érosion alvéolaire et de 'taffonis'. *C.r. somm. & Bull. Soc. Geol. Fr.* 6(6): 135-136.

Denham, D., Alexander, L.T. and Worotnicki, G., 1979. Stresses in the Australian crust: evidence from earthquakes and *in-situ,* stress measurements. *B.M.R. J. Austr. Geol. & Geophys.* 4: 289-295.

Dorman, F.H., 1966. Australian Tertiary palaeotemperatures. *J. Geol.* 74: 49-61.

Dorn, R.I. and Oberlander, T.M., 1981. Microbial origin of desert varnish. *Science* 213: 1245-1247.

Dragovich, D.J., 1964. *Cavernous Weathering of Granitic and Some Gneissic Rocks in southern South Australia.* Unpubl. M.A. thesis, University of Adelaide.

Dragovich, D.J., 1966. Granite lapiés at Remarkable Rocks, South Australia. *Rev. Géomorph. Dynam.* 18: 8-16

Dragovich, D.J., 1967. Flaking, a weathering process operating on cavernous rock surfaces. *Geol. Soc. Amer. Bull.* 78: 801-804.

Dragovich, D.J., 1969. The origin of cavernous surfaces (tafoni) in granitic rocks of southern South Australia. *Z. Geomorph.* 13: 163-181.

Drake, F.E., 1859. Artificial origin of rock-basins. *Geologist,* 2: 368-371.

Dresch, J., 1949. Sur les pédiments en Afrique Mediterranéene et tropicale. *C.r. Inter. Géogr. Congr. (Lisbon)* 2: 19-28.

Dresch, J., 1957. Pédiments et glacis d'érosion, pediplaines et inselbergs. *L'Inf. Géogr.* 21: 183-196.

Duffaut, P., 1957. Sur la genèse des 'boules' de certains granites. *C.r. somm. & Bull. Soc. Géol. France* 7: 139-141.

Dumanowski, B., 1960. Comment on origin of depressions surrounding granite massifs in the eastern desert of Egypt. *Bull. Acad. Pol. Sci.* 8: 305-312.

Dumanowski, B., 1964. Problem of the development of slopes in granitoids. *Z. Geomorph. Supp.-Band* 5: 30-40.

Dumanowski, B., 1968. Influence of petrological differentiation of granitoids on land forms. *Geogr. Pol.* 7: 93-98.

Durand Delga M., 1978. *Corse.* Masson, Paris.

Dury, G.H. and Langford-Smith, T., 1964. The use of the term peneplain in descriptions of Australian landscapes. *Austr. J. Sci.* 27: 171-175.

Dzulinski, S.T., and Kotarba, A., 1979. Solution pans and their bearing on the development of pediments and tors in granite. *Z. Geomorph.* 23: 172-191.

Eggler, D.H., Larson, E.E. and Bradley, W.C., 1969. Granites, gneisses and the Sherman erosion surface, southern Laramie Range, Colorado-Wyoming. *Amer. J. Sci.* 267: 510-522.

Eggleton, R.A. and Buseck, P.R., 1980. High resolution electron microscopy of feldspar weathering. *Clays and Clay Minerals* 28: 173-178.

Eiseley, L., 1961. *Darwin's Century.* Doubleday, New York.

Elders, W.A., 1963. On the form and emplacement of the Herefoss granite. *Norsk. Geol. Undersökelse* 214A.

Embleton, B.J.J., 1973. The palaeolatitude of Australia through Phanerozoic time. *J. geol. Soc. Aust.* 19: 475-482.

Emerson, B.K., 1898. Geology of Old Hampshire County, Mass. *U.S. geol. Surv. Mon.* 39.

Emery, K.O., 1944. Bushfires and rock exfoliation. *Amer. J. Sci.* 242: 506-508.

Engel, C.G., and Sharp, R.P., 1958. Chemical data on desert varnish. *Geol. Soc. Amer. Bull.* 69: 487-518.

Eskola, P., 1949. The problem of mantled gneiss domes. *Quart. J. geol. Soc. London,* 104: 461-476.

Evans, I.S., 1969. Salt crystallization and rock weathering; a review. *Rev. Géomorph. Dynam.* 19: 155-177.

Fairbridge, R.W. and Finkl, C.W., 1978. Geomorphic analysis of the rifted cratonic margins of Western Australia. *Z. Geomorph.* 22: 369-389.

Falconer, J.D., 1911. *The Geology and Geography of Northern Nigeria.* Macmillan, London.

Farmin, R., 1937. Hypogene exfoliation of rock masses. *J. Geol.* 45: 625-635.

Feininger, T., 1969. Pseudokarst on quartz diorite, Columbia. *Z. Geomorph.* 13: 287-286.

Finlayson, B., 1981. Underground streams on acid igneous rocks in Victoria. *Helictite* 19: 5-14.

Fisher, O., 1872. On cirques and taluses. *Geol. Mag.* 8: 10-12.

Ford, T.D., 1962. The dolomite tors of Derbyshire. *E. Midl. geogr.* 3: 148-153.

Foureau, F., 1905. *Documents scientifiques de la Mission Saharienne.* Masson, Paris.

Francis, W.D., 1921. The origin of black coatings of iron and manganese oxides on rocks. *Proc. Royal Soc. Qld.* 32: 110-116.

Frenzel, G., 1965. Studies on Mediterranean tafoni. *Neue Jahr. f. Geol. Min. Pal.* 122: 313-323.

Fry, E.J., 1924. A suggested explanation of the lithophytic lichens on rocks (shale). *Ann. Bot.* 38: 175-196.

Fry, E.J., 1926. The mechanical action of corticolous lichens. *Ann. Bot.* 40: 397-417.

Fry, E.J., 1927. The mechanical action of crustaceous lichens, on substrata of shale, schist, gneiss, limestone and obsidian. *Ann. Bot.* 41: 437-460.

Fuge, R., 1979. Water-soluble chlorine in granitic rocks. *Chem. Geol.* 25: 169-174.

Gage, M., 1966. Franz Josef Glacier. *Ice,* 20: 26-27.

Galli-Olivier, C., 1967. Pediplain in northern Chile and the Andean uplift. *Science* 158: 653-655.

Gee, R.D. and Groves, D.I., 1971. Structural features and mode of emplacement of part of the Blue Tier Batholith in northeastern Tasmania. *J. geol. Soc. Aust.* 18: 41-56.

Geikie, A., 1886. *A Case-Book of Geology.* Macmillan, London.

Geological Society of Australia, 1971. *Tectonic Map of Australia and New Guinea 1:5 000 000* Sydney.

Gerrard, A.J.W., 1974. The geomorphological importance of jointing in the Dartmoor granites, pp. 39-51. In: E.H. Brown and R.S. Waters (Editors) *Progress in Geomorphology.* Inst. Brit. Geogr. Spec. Publ. 7, London.

Gerth, H., 1955. *Der Geologische Bau der Südamerikanischen Kordillere.* Borntraeger, Berlin.

Gifkins, R.C., 1959. Mechanisms of intergranular failure at elevated temperatures. In: *Conference on Fracture,* NAS/NRC Swampscott, Mass.

Gifkins, R.C., 1965. Intergranular creep fracture, pp. 44-61. In: C.J. Osborn (Editor) *Fracture* (Proc. First Tewkesbury Symp., Melbourne, 1963). Univ. Melbourne, Melbourne.

Gilbert, G.K., 1904. Domes and dome structures of the High Sierra. *Geol. Soc. Amer. Bull.* 15: 29-36.

Giles, E., 1889. *Australia Twice Traversed* Vols. 1 and 2, Sampson Low, Marston, Searle and Rivington, London (Australiana Facsimile Ed. No. 13, Lib. Board, S.A., Adelaide 1964).

Gilluly, J., 1949. Distribution of mountain building in geologic time. *Geol. Soc. Amer. Bull.* 60: 561-590.

Gilluly, J., 1955. Geologic contrasts between continents and ocean basins. *Geol. Soc. Amer. Spec. Paper* 62: 7-18.

Godard, A., 1977. *Pays et Paysages du Granite.* Presses Univ. France, Paris.

Godard, A., 1979. Reconnaissance dans l'éxtremité nord du Labrador et du Nouveau Québec: Contribution a l'étude géomorphologique des socles des milieux froids. *Rev. Géomorph. Dynam.* 28: 125-142.

Godard, A., *et al.*, 1972. Quelques enseignements apportés par le Massif Central Français dans l'étude géomorphologique des socles crystallins. *Rev. Géogr. Phys. Géol. Dynam.* 14: 265-296.

Goldich, S.A., 1938. A study in rock-weathering. *J. Geol.* 46: 17-58.

Gosse, W.C., 1874. Report and diary of Mr. W.C. Gosse's central and western exploring expedition, 1873. *S.A. Parl. Paper* 48.

Goudie, A., 1971. Climate, weathering, crust formation, dunes and features of the central Namibia Desert near Gobabeb, South West Africa. *Madoque* 2: 15-31.

Goudie, A., 1974. Further experimental investigation of rock weathering by salt and other mechanical processes. *Z. Geomorph. Suppl.-Band.* 21: 1-12.

Grawe, O.R., 1936. Ice as an agent of rock weathering: a discussion. *J. Geol.* 44: 173-182.

Griggs, D.T., 1936. The factor of fatigue in rock exfoliation. *J. Geol.* 44: 783-796.

Hack, J.T., 1966. Circular patterns and exfoliation in crystalline terrains, Grandfather Mountain area, North Carolina. *Geol. Soc. Amer. Bull.* 77: 975-986.

Hake, B.E., 1928. Scarps of the southwestern Sierra Nevada, California. *Geol. Soc. Amer. Bull.* 39: 1017-1030.

Hale, M.E., 1967. *The Biology of Lichens.* Arnold, London.

Hambleton-Jones, B.B., 1976. *The Geology and Geochemistry of some Epigenetic Uranium deposits near the Swakop River, South West Africa.* Unpubl. D.Sc. thesis, Univ. Pretoria.

Hamilton, W. and Myers, W.B., 1967. The nature of batholiths. *U.S. geol. Surv. Prof. Paper* 554-C.

Handley, J.R.F., 1952. The geomorphology of the Nzega area of Tanganyika with special reference to the formation of granite tors. *C.r. Congr. géol. Inter.* (Algiers) 21: 201-210.

Hankar-Urban, A., 1906. Note sur les mouvements spontanés des roches dans les carrières. *Bull. Soc. Belge Pal. Hydrol.* 19: 527-540.

Harland, W.B., 1957. Exfoliation joints and ice action. *J. Glac.* 3: 8-10.

Harris, G.F., 1888. *Granite and Our Granite Industries.* Crosby and Lockwood, London.

Harriss, R.C. and Adams, J.A.S., 1966. Geochemical and mineralogical studies on the weathering of granitic rocks. *Amer. J. Sci.* 264: 146-173.

Hartt, C.F., 1870. *Geology and Physical Geography of Brazil.* Trübner, London.

Hassenfratz, J-H., 1791. Sur l'arrangement de plusieurs gros blocs de differentes pierres que l'on observe dans les montagnes. *Ann. Chimie,* 11: 95-107.

Hedges, J., 1969. Opferkessel. *Z. Geomorph.* 13: 22-55.

Hedges, J., 1978. Karst caves in silicate rocks. *D.C. Speleo.,* 3-4.

Helbig, K., 1940. Die Insel Bangka. *Deutsch Geogr. Baltter,* 43: 133-207.

Hellstrom, B., 1941. Nagra lakttagelser över Vitting, Erosion Och Slambildning I Malaya ich Australien. *Geogr. Ann.* 23: 102-24.

Herrmann, L.A., 1957. Geology of the Stone Mountain - Lithonia district, Georgia. *Georgia Geol. Surv. Bull.* 61.

Herrmann, O., 1895. Technische Verwerthung der Lausitzer Granite. *Z. prakt. Geologie,* pp. 433-444.

Hills, E.S., 1940. *Physiography of Victoria: an introduction to geomorphology.* Whitcombe and Tombs, Melbourne.

Hills, E.S., 1949. Shore Platforms. *Geol. Mag.* 86: 137-152.

Hills, E.S., 1955. Die landoberfläche Australiens. *Die Erde* 7: 195-205.

Hills, E.S., 1963. *Elements of Structural Geology.* Methuen, London.

Hills, E.S., 1971. A study of cliffy coastal profiles based on examples in Victoria, Australia. *Z. Geomorph.* 15: 137-180.

Hills, E.S., 1975. *Physiography of Victoria.* Whitcombe and Tombs, Melbourne

Hitchcock, C.B., 1947. The Orinoco-Ventuari region, Venezuela. *Geogr. Rev.* 37: 525-566.

Hodgson, J.H., 1964. *Earthquakes and Earth Structure.* Prentice-Hall, New Jersey.

Hoek, E., 1968. Brittle failure of rock. Chapter 4. In: K.G. Stagg and O.C. Zienkiewicz (Editors) *Rock Mechanics in Engineering Practice*. Wiley, London.

Hoek, E., and Bieniawski, Z.T., 1965. Brittle fracture propagation in rocks under compression. *Inter. J. Fracture Mech*. 1: 137-155.

Högbom, B., 1912. Wüstenerscheinungen auf Spitzbergen. *Bull. geol. Instn. Univ. Uppsala* 11: 242-251.

Holmes, A., 1918. The Pre-Cambrian and associated rocks of the District of Mozambique. *Quart. J. geol. Soc. London* 74: 31-97.

Holmes, A., 1965. *Principles of Physical Geology*. Nelson, Edinburgh.

Holmes, A. and Wray, D.A., 1912. Outlines of the geology of Mozambique. *Geol. Mag*. 9: 412-417.

Holmes, A. and Wray, D.A., 1913. Mozambique - a geographical study. *Geogrl. J*. 42: 143-152.

Hooke, R. LeB., Chang, Houng-YI and Weiblen, P.W., 1969. Desert varnish: an electron probe study. *J. Geol*. 77: 275-288.

Howard, A.D., 1942. Pediment passes and the pediment problem. *J. Geomorph*. 5: 3-31, 95-136.

Howchin, W., 1918. *An Introduction to Geology, (Physiographic and Structural), From the Australian Standpoint*. Govt. Printer, Adelaide.

Humboldt, A. von, 1852. *Personal Narrative* (Trans. and Ed. T. Ross), Vol. 2, Bohn, London.

Hume, W.F., 1925. *Geology of Egypt, Vol. I*. Govt. Printer, Cairo.

Hunt, C.B., 1953. Geology and geography of the Henry Mountains region, Utah. *U.S. geol. Surv. Prof. Paper* 228.

Hurault, J., 1963. Recherches sur les inselbergs granitiques nus en Guyane Française. *Rev. Géomorph. Dynam*. 14: 49-61.

Hutton, J., 1795. *Theory of the Earth*. Creech, Edinburgh.

Hutton, J.T., 1981. Clays and bricks of the penal settlements at Port Arthur and Maria Island, Tasmania. *Proc. Royal Soc. Tas*. 115: 153-161.

Hutton, J.T., Lindsay, D.S. and Twidale, C.R., 1977. The weathering of norite at Black Hill, South Australia. *J. geol. Soc. Aust*. 24: 37-50.

Ingles, O.G., Lee, I.K. and Neil, R.C., 1972. A new manifestation of stress history. *Univciv. Rept. L-90. Univ. of N.S.W., (Sydney)*.

Isaacson, E. de St. Q., 1957. Research into the rock burst problem on the Kolar Goldfield. *Mine Quarry Engng*. 23: 520-526.

Isherwood, D. and Street, A., 1976. Biotite-induced grussification of the Boulder Creek Granodiorite, Boulder County, Colorado. *Geol. Soc. Amer. Bull*. 87: 366-370.

Jack, R.L., 1915. Geology and prospects of the region to the south of the Musgrave Ranges and the geology of the western part of the Great Artesian Basin. *Mines Dept. S. Aust. Bull*. 5.

Jaeger, J.C., 1965. Fracture of rocks, pp. 268-283. In: C.J. Osborn (Editor) *Fracture* (Proc. First Tewkesbury Symp., Melbourne, 1963). Univ. Melb., Melbourne.

Jahn, A., 1974. Granite tors in the Sudeten Mountains, pp. 53-61. In: E.H. Brown and R.S. Waters (Editors) *Progress in Geomorphology*. Inst. Brit. geogr. (Spec. Publ. 7), London.

Jahns, R.H., 1943. Sheet structure in granites: its origin and use as a measure of glacial erosion in New England. *J. Geol*. 51: 71-98.

Jameson, R., 1798. *An outline of the Mineralogy of the Shetland Islands and of the Islands of Arran*. Creech, Edinburgh.

Jameson, R., 1820. Mineralogy. In: D. Brewster (Editor) *Edinburgh Encyclopaedia* 14 (II), Blackwood, Edinburgh. Chapters 3-5 inclusive.

Jeje, L.K., 1973. Inselbergs' evolution in a humid tropical environment: the example of south-western Nigeria. *Z. Geomorph*. 17: 194-225.

Jennings, J.N., 1976. A test of the importance of cliff-foot caves in tower karst development. *Z. Geomorph. Suppl.-Band* 26: 92-97.

Jennings, J.N., 1978. Genetic variety in A-tents and related features. *Austr. geogr*. 14: 34-38. *Addendum*, p.62.

Jennings, J.N. and Sweeting, M.M., 1961. Caliche pseudo-anticlines in the Fitzroy Basin, Western Australia. *Amer. J. Sci*. 259: 635-639.

348

Jennings, J.N. and Twidale, C.R., 1971. Origin and implications of the A-tent, a minor granite landform. *Austr. geogr. Stud.* 9: 41-53.

Jessen, O., 1936. *Reisen und Forschungen in Angola*. Reimer, Berlin.

Johnson, A.R.M., 1974. Cavernous weathering at Berowra, N.S.W. *Austr. geogr.* 12: 531-553.

Johnson, R.J., 1927. Polygonal weathering in igneous and sedimentary rocks. *Amer. J. Sci.* 13: 440-444.

Johnston, J.H., 1973. Salt weathering processes in McMurdo dry valley regions of South Victoria Land, Antarctica. *N.Z. J. Geol. Geophys.* 16: 221-224.

Joly, F., 1949. Pédiments et glacis d'érosion dans le SE du Maroc. *C.r. Congr. Inter. Géogr. (Lisbon)* 2: 110-125.

Joly, F., 1952. Erosion en surface et d'érosion liniaire dans le modelé prédesertique, pp. 255-267. In: *Lab. geogr. de Rennes, Vol. Jubil. offert à E de Martonne*.

Joly, J., 1901. Expériences sur la denudation par dissolution dans l'eau douce et l'eau de mer. *C.r. Congr. Géol. Inter. (Paris) VIII (II)*, pp. 774-784.

Jones, T.R., 1859. Notes on some granite tors. *Geologist* 2: 301-312.

Jutson, J.T., 1914. An outline of the physiographical geology (physiography) of Western Australia. *Geol. Surv. W. Aust. Bull.* 61.

Jutson, J.T., 1917. Erosion and the resulting landforms in sub-arid Western Australia, including the origin and growth of the dry lakes. *Geogrl. J.* 50: 418-437.

Jutson, J.T., 1934. The physiography (geomorphology) of Western Australia. *Geol. Surv. W. Aust. Bull.* 95.

Jutson, J.T., 1940. The shore platforms of Mt. Martha, Port Phillip Bay, Victoria, Australia. *Proc. Royal Soc. Vict.* 52: 164-176.

Kastning, E.H., 1976. Granitic karst and pseudokarst, Llano County, Texas, with special reference to Enchanted Rock Cave. *Proc. 1976 N.S.S. Annual Convention*, pp. 43-45.

Keller, W.D. and Frederickson, A.F., 1952. Role of plants and colloidal clays in the mechanism of weathering. *Amer. J. Sci.* 250: 594-608.

Kelly, W.C. and Zumberge, J.H., 1961. Weathering of a quartz diorite at Marble Point, McMurdo Sound, Antarctica. *J. Geol.* 69: 433-446.

Kemp, E.M., 1978. Tertiary climatic evolution and vegetation history in the southeast Indian Ocean region. *Palaeogeog. Palaeoclim. & Palaeoecol.* 24: 169-208.

Kessler, D.W., Insley, H. and Sligh, W.H. 1940. Physical, mineralogical and durability studies on the building and monumental granites of the United States. *J. Res. natnl. Bur. Stand.* 24: 161-206.

Keyes, C.R., 1912. Deflative scheme of the geologic cycle in an arid climate. *Geol. Soc. Amer. Bull.* 23: 537-562.

Kiersch, G.A., 1964. Vaiont Reservoir disaster. *Civ. Engng.* 34: 32-39.

Kieslinger, A., 1932. *Zerstörungen an Steinbauten, ihre Ursachen und ihre Abwehr*. Deuticke, Leipzig.

Kieslinger, A., 1960. Residual stress and relaxation in rocks. *Inter. Geol. Cong. (Copenhagen), (21)* 18: 270-276.

King, L.C., 1942. *South African Scenery: a Textbook of Geomorphology*. Oliver and Boyd, Edinburgh.

King, L.C., 1949a. A theory of bornhardts. *Geogrl. J.* 112: 83-87.

King, L.C., 1949b. The pediment landform: some current problems. *Geol. Mag.* 86: 245-250.

King, L.C., 1950. A study of the world's plainlands. *Quart. J. geol. Soc. London* 106: 101-131.

King, L.C., 1953. Canons of landscape evolution. *Geol. Soc. Amer. Bull.* 64: 721-752.

King, L.C., 1957. The uniformitarian nature of hillslopes. *Trans. geol. Soc. Edinburgh*, 17: 81-102.

King, L.C., 1962. *Morphology of the Earth*. Oliver and Boyd, Edinburgh.

King, L.C., 1966. The origin of bornhardts. *Z. Geomorph.* 10: 97-98.

Kingsmill, T.W., 1862. Notes on the geology of the east coast of China. *J.R. geol. Soc. Ireland*, 10: 1-6.

Kinzl, H. and Schneider, E., 1950. *Cordillera Blanca, Perú*. Wagner, Innsbruck.

Klaer, W., 1956. Verwitterungsformen in Granit auf Korsika. *Pet. geogr. Mitt. Ergänzungsheft* 261.

Klaer, W., 1957. Verkarstungserscheinungen in Silikatgestein. *Abb. Geogr. Inst. Freien Univ. Berlin*, 5: 21-27.

Klaer, W., 1973. Untersuchungen zur klimagenetischen Geomorphologie im Granit auf Korsika. *Geogr. Z.* 33: 247-260.

Kranck, E.H., 1957. On folding movements in the zone of the basement. *Geol. Rdsch.* 46: 261-282.

Krauskopf, K.B., 1953. Tungsten deposits of Modera, Fresno, and Tulare counties, California. *Calif. Div. Mines Special Rept.* 35.

Krieg, G.W., 1972. *Everard*. Map Sheet SG53-13 Zone 4, Geol. Surv. S. Aust.

Kukal, Z. and Adnan, B.A.N., 1970. Scratch circles on the Pliocene sandstones in central Iraq. *Z. Geomorph.* 14: 329:334.

Kvelberg, I. and Popoff, B., 1937. Die Tafoni - Verwitterungserscheinung. *Univ. Latviensis (Riga), Kimijas Fac. Acta*, 4, (6): 129-368.

Kwaad, F.J.P.M., 1970. Experiments on the granular disintegration of granite by salt action. *Univ. Amsterdam Rysisch Geogr. Boden Lab. Publ.* 16.

Lagasquie, J.T., 1978. Relations entre les modelés d'érosion différentielle et la structure de quelques ensembles de granitoides des Pyrénées Centrales et orientales. *Rev. Géog. Phys. Géol. Dynam.* 20: 219-234.

Lahee, F.H., 1909. Theory and hypothesis in geology. *Science* 30: 562-563.

Lamego, A.R., 1938. Escarpas do Rio de Janeiro. *Depart. Nac. Producao Mineral (Brasil) Serv. Geol. Mineral. Bol.* 93.

Lane, E.W., et al., 1947. Report of the subcommittee on sedimentary terminology. *Trans. Amer. geophys. Union*. 1947: 936-938.

Lapparent, A. de, 1907. *Leçons de Géographie Physique*. Masson (3rd edition), Paris.

Lapworth, C. and Watts, W.W., 1894. The geology of south Shropshire. *Proc. geol. Ass. (London)* 13: 297-355.

Larsen, E.S., 1948. Batholith and associated rocks of Corona, Elsinore and San Luis Rey Quadrangles, Southern California. *Geol. Soc. Amer. Mem.* 29: 113-119.

Laurie, A.P., 1925. Stone decay and the preservation of buildings. *J. Soc. Chem. Ind.* 44B: 86-92.

Lautansach, H., 1950. Granitische Abtragungsformen auf der Iberischen Halbinsel und in Korea, ein Vergleich. *C.r. Cong. Inter. Géogr. (Lisbon)* 1946: 270-296.

Lawson, A.C., 1915. The epigene profiles of the desert. *Univ. Calif. (Berkeley) Publ. Geology*, 9: 23-40.

Le Conte, J.N., 1873. On some of the ancient glaciers of the Sierras. *Amer. J. Sci. & Arts* 5: 325-342.

Lee, W.T., 1922. Peneplains of the Front Range and Rocky Mountain National Park, Colorado. *U.S. geol. Surv. Bull.* 730-A.

Leeman, E.F., 1962. Rock bursts in South African gold mines. *New Scientist* 16: 79-82.

Lehmann, H., 1954. Der Tropische Kegelkarst auf der Grossen Antillen. *Erdkunde* 8: 130-139.

Leigh, C.H., 1967. *Some aspects of the geomorphology of granitic landforms and landscapes on parts of the New England Tableland, New South Wales*. Unpubl. Ph.D thesis, Univ. of New England.

Leigh, C.H., 1970. Australian landform example No.16: Tors of subsurface origin. *Austr. geogr.* 11: 288-290.

Leonard, R.J., 1929. Polygonal cracking in granite. *Amer. J. Sci.* 18: 487-492.

Lester, J.G., 1938. Geology of the region round Stone Mountain, Georgia. *Univ. Colorado Studies*, Ser. A, 26: 88-91.

Lewis, W.V., 1954. Pressure release and glacial erosion. *J. Glaciol.* 2: 417-422.

Linton, D.L., 1952. The significance of tors in glaciated lands. *Inter. Géogr. Union Proc. 8th Gen. Assemb. (Lisbon) 17th Inter. Congr.*, pp. 354-357.

Linton, D.L., 1955. The problem of tors. *Geogrl. J.* 121: 470-487.

Linton, D.L., 1957. The everlasting hills. *Adv. Sci.* 14: 58-67.

Linton, D.L., 1963. The forms of glacial erosion. *Trans. Inst. Brit. Geogr.* 33: 1-28.

Linton, D.L., 1964. The origin of the Pennine tors - an essay in analysis. Z. Geomorph. 8: 5-24.

Lister, L.A., 1973. The microgeomorphology of granite hills in north-eastern Rhodesia. Geol. Soc. S. Afr. Spec. Publ. 3: 157-161.

Ljunger, E., 1930. Spaltentektonik und Morphologie der Schwedischen Skagerrak-Küste. Bull. geol. Instn. Univ. Uppsala, 21: 1-478.

Lobeck, A.K., 1939. Geomorphology. McGraw-Hill, New York.

Logan, J.R., 1848. Sketch of the physical geography and geology of the Malay Peninsula. J. Indian Archipelago & East Asia (1), 2: 83-138.

Logan, J.R., 1849. The rocks of Palo Ubin. Verh. Genootsch. van Kunst. Wetenschappen (Batavia) 22: 3-43.

Logan, J.R., 1851. Notices of the geology of the Straits of Singapore. Quart. J. geol. Soc. London 7: 310-344.

Logan, R.F., 1960. The Central Namib Desert, South West Africa. Nat. Acad. Sci. Nat. Research Council Pub. 758, p.162.

Loudermilk, J.D., 1931. On the origins of desert varnish. Amer. J. Sci. 21: 51-66.

Loughnan, F.C., 1969. Chemical Weathering of the Silicate Minerals. Elsevier, London.

Louis, H., 1960. Allgemeine Geomorphologie. de Gruyter, Berlin.

Ludbrook, N.H., 1969. Tertiary Period, pp. 172-203. In: L.W. Parkin (Editor) Handbook of South Australian Geology. Geol. Surv. S. Aust., Adelaide.

Mabbutt, J.A., 1952. A study of granite relief from South West Africa. Geol. Mag. 89: 87-96.

Mabbutt, J.A., 1955. Erosion surfaces in Namaqualand and the ages of surface deposits in the south-western Kalahari. Trans. geol. Soc. S. Africa 58: 13-30.

Mabbutt, J.A., 1961a. 'Basal surface' or 'weathering front'. Proc. geol. Ass. (London), 72: 357-358.

Mabbutt, J.A., 1961b. A stripped land surface in Western Australia. Trans. & Papers Inst. Brit. geogr. 29: 101-114.

Mabbutt, J.A., 1965. The weathered land surface in central Australia. Z. Geomorph. 9: 82-114.

Mabbutt, J.A., 1966. Mantle-controlled planation of pediments. Amer. J. Sci. 264: 78-91.

Mabbutt, J.A., 1967. Denudation chronology in central Australia, pp. 144-181. In: J.N. Jennings and J.A. Mabbutt (Editors) Landform Studies from Australia and New Guinea. Aust. Natl. Univ. Press, Canberra.

MacCulloch, J., 1814. On the granite tors of Cornwall. Trans. geol. Soc. 2: 66-78.

MacGregor, A.M., 1951. Some milestones in the Precambrian of Southern Rhodesia. Trans. geol. Soc. S. Afr. 54: xxvii-lxxi.

Mackin, J.H., 1937. Erosional history of the Big Horn Basin, Wyoming. Geol. Soc. Amer. Bull. 48: 815-860.

Mackin, J.H., 1970. Origin of pediments in the western United States, pp. 85-105. In: M. Pecsi (Editor) Problems of Relief Planation. Akadémiai Kiado, Budapest.

Maclaren, M., 1912. Notes on desert-water in Western Australia, 'Gnamma' holes and 'night-wells'. Geol. Mag. 9: 301-304.

MacMahon, C.A., 1893. Notes on Dartmoor. Quart. J. geol. Soc. London, 49: 385-397.

Marti Boni C. and Vidal Romani, J.R., 1981. Datos para la Comparación del micromodelado en dos macizos de granitoides peninsulares. Cuad. Lab. Xeol. Laxe 2: 265-270.

Martini, I.P., 1978. Tafoni weathering with examples from Tuscany, Italy. Z. Geomorph. 22: 44-67.

Martonne, E. de, 1925. Traité de Géographie Physique. II. Colin, Paris.

Martonne, E. de, 1951. Traité de Géographie Physique. II Le Relief du Sol. Colin, Paris.

Mason, B., 1966. Principles of Geochemistry. (3rd edition). Wiley, New York.

Matschinski, M., 1954. Quelques considerations sur la théorie mathématique des taffoni. Acad. Naz. Lincei rend. (Ser. 8), 16: 632-363, 731-734.

Matthes, F.E., 1930. Geologic history of the Yosemite Valley. *U.S. geol..Surv. Prof. Paper 160.*

Matthes, F.E., 1937. The geologic history of Mount Whitney. *Sierra Club Bull.* 22: 1-18.

McGee, W.J., 1897. Sheetflood erosion. *Geol. Soc. Amer. Bull.* 8: 87-112.

Mennell, F.P., 1904. Some aspects of the Matopos. I. Geological and physical features. *Proc. Rhod. Sci. Assoc.* 4: 72-76.

Mensching, H., 1958. Glacis-Fussfläche-Pediment. *Z. Geomorph.* 2: 165-186.

Merrill, G.P., 1898. *Treatise on Rocks, Weathering and Soils.* Macmillan, London.

Meunier, A. and Velde, B., 1976. Mineral reactions at grain contacts in early stages of granite weathering. *Clay Minerals* 11: 235-240.

Michel-Mainguet, M., 1972. *Le Modelé des Grès.* Inst. Géogr. Natl., Paris.

Miller, W.J., 1940. *Elements of Geology.* van Nostrand, New York.

Mortensen, H., 1933. Die 'Salzsprengung' und Ihre Bedeutung für die regional klimatische Gliederung der Wüsten. *Pet. geogr. Mitt.* 79: 130-135.

Moss, A.J., 1973. Fatigue effects in quartz sand grains. *Sediment. Geol.,* 10: 239-247.

Moss, A.J. and Green, P., 1975. Sand and silt grains: predetermination of their formation and properties by microfractures in quartz. *J. geol. Soc. Aust.* 22, 4: 485-495.

Moss, A.J., Green, P. and Hutka, J., 1981. Static breakage of granitic detritus by ice and water in comparison with breakage by flowing water. *Sedimentology* 28: 261-272.

Moss, A.J., Walker, P.H. and Hutka, J., 1973. Fragmentation of granitic quartz in water. *Sedimentology* 20: 489-511.

Moye, D.G., 1958. *Rock mechanics in the interpretation and construction of T.1 underground power station, Snowy Mountains, Australia.* Paper presented at A.G.M. Geol. Soc. Amer.; symposium sponsored by Am. Soc. Civil Eng. & Geol. Amer. J. Comm. on Engng. Geol., St. Louis, 6-8 Nov. 1958.

Nesbitt, H.W., 1979. Mobility and fractionation of rare earth elements during weathering of a granodiorite. *Nature* 279 (5710): 206-210.

Nesbitt, H.W., Markovics, G. and Price, R.C., 1980. Chemical processes affecting alkalis and alkaline earths during continental weathering. *Geochem. Cosmochim. Acta* 44: 1659-1666.

Niles, W.H., 1872. Some interesting phenomena observed in quarrying. *Proc. Boston Nat. Hist. Soc.* 14: 80-87; 16: 41-43.

Noldart, A.J. and Wyatt, J.D., 1962. The geology of partion of the Pilbara Goldfield. *Geol. Surv. W. Aust. Bull.* 115.

Oberlander, T.M., 1972. Morphogenesis of granitic boulder slopes in the Mojave Desert, California. *J. Geol.* 80: 1-20.

Obst, E., 1923. Das abflusslose Rumpfschollenland in nordöstlichen Deutsch-Östafrica. *Mitt. geogr. Gesell. Hamb.* 35.

Ollier, C.D., 1960. The inselbergs of Uganda. *Z. Geomorph.* 4: 43-52.

Ollier, C.D., 1963. Insolation weathering: examples from central Australia. *Amer. J. Sci.* 261: 376-381.

Ollier, C.D., 1965. Some features of granite weathering in Australia. *Z. Geomorph.* 9: 285-304.

Ollier, C.D., 1967. Spheroidal weathering, exfoliation and constant volume alteration. *Z. Geomorph.* 11: 103-108.

Ollier, C.D., 1971. Causes of spheroidal weathering. *Earth Sci. Rev.* 7: 127-141.

Ollier, C.D., and Pain, C.F., 1980. Actively rising surficial gneiss domes in Papua New Guinea. *J. geol. Soc. Aust.* 27: 33-44.

Ollier, C.D., and Tuddenham, W.G., 1962. Inselbergs of central Australia. *Z. Geomorph.* 5: 257-276.

Öpik, A.A., 1961. The geology and palaeontology of the headwaters of the Burke River, Queensland. *Bur. Mines. Resour. Geol. Geophys. Aust. Bull.* 53.

Orme, A.R., 1964. The geomorphology of southern Dartmoor and the adjacent area, pp. 31-72. In: I.G. Simmons (Editor) *Dartmoor Essays.* Devon. Assoc. Adv. Sci. Lit. & Art.

Ormerod, G.W., 1859. On the rock basins in the granite of the Dartmoor District, Devonshire. *Quart. J. geol. Soc. London* 15: 16-29.

Ormerod, G.W., 1869. On the results arising from the bedding, joints and spheroidal structure of the granite on the eastern side of Dartmoor, Devonshire. *Quart. J. geol. Soc. London* 25: 273-280.

Osborn, F.F., 1935. Rift, grain and hardway in some pre-Cambrian granites, Quebec. *Econ. Geol.* 30, 5: 540-551.

Paige, S., 1912. Rock-cut surfaces in the desert ranges. *J. Geol.* 20: 442-540.

Pain, C.F. and Ollier, C.D., 1981. Geomorphology of a Pliocene granite in Papua New Guinea. *Z. Geomorph.* 25: 249-258.

Pallister, J.W., 1960. Erosion cycles and associated surfaces of Mengo District, Buganda. *Overseas Geol. & Min. Res.* 8: 26-36.

Palmer, J. and Nielson, R.A., 1962. The origin of granite tors on Dartmoor, Devonshire. *Proc. Yorks. geol. Soc.* 33: 315-340.

Palmer, J. and Radley, J., 1961. Gritstone tors of the English Pennines. *Z. Geomorph.* 5: 37-52.

Parslow, G.R., 1968. The Physical and structural features of the Cairnsmore of Fleet granite and its aureole. *Scot. J. Geol.* 4: 91-108.

Passarge, S., 1895. *Adamaua*. Reimer, Berlin.

Passarge, S., 1923. Die Inselberglandschaft der Massaisteppe. *Pet. geogr. Mitt.* 69: 205-209.

Passarge, S., 1928. *Panoramen Afrikanischer Inselberglandschaft*. Reimer, Berlin.

Peel, R.F., 1939. The Gilf Kebir. *Geogrl. J.* 93: 295-307.

Peel, R.F., 1941. Denudation landforms of the central Libyan Desert. *J. Geomorph.* 4: 3-23.

Peel, R.F., 1960. Some aspects of desert geomorphology. *Geogr.* 45: 241-262.

Penck, A., 1894. *Morphologie der Erdoberfläche*. Engelhorns, Stuttgart.

Penck, W., 1924. *Die Morphologische Analyse*. Engelhorns, Stuttgart.

Penck, W., 1953. *Morphological Analysis of Landforms*. (Transl. H. Czeck and K.C. Boswell). Macmillan, London.

Perry, R.S. and Adams, J.B., 1978. Desert varnish: evidence for cyclic deposition of manganese. *Nature* (276): 489-491.

Peterson, J.A., 1975. An A-tent from plateau Labrador. *Aust. geogr. Stud.* 13: 195-199.

Pettijohn, F.J., 1957. *Sedimentary Rocks*. Harper, New York.

Pirsson, L.V. and Schuchert, C., 1915. *A Text-Book of Geology*. Wiley, New York.

Playfair, J., 1802. *Illustrations of the Huttonian Theory of the Earth*. Creech, Edinburgh.

Plenderleith, H.J., 1956. *The Conservation of Antiquities and Works of Art*. Oxford U.P., London.

Potts, A.S., 1970. Frost action: some experimental data. *Trans. Inst. Brit. geogr.* 49: 109-124.

Pouyllau, D. and Seurin, M., 1982. Geomorphologie et pseudo-karsts sur grès et quartzites du Roraima dans la region de la Gran Sabana (sud-est du Vénézuela). *Trav. Doc. Géogr. Trop.* 44.

Powers, W.E., 1936. The evidences of wind abrasion. *J. Geol.* 44: 214-219.

Price, N.J., 1966. *Fault and Joint Development in Brittle and Semi-brittle Rock*. Pergamon, London.

Price, W.A., 1925. Caliche and pseudo-anticlines. *Bull. Assoc. Amer. Petrol. Geol. Sci.* 9: 1009-1017.

Pugh, J.C., 1956. Fringing pediments and marginal depressions in the inselberg landscape of Nigeria. *Trans. & Papers Inst. Brit. geogr.* 22: 15-31.

Pumpelly, R., 1879. The relation of secular rock disintegration to loess, glacial drift and rock basins. *Amer. J. Sci.* 17: 133-144.

Rahn, P.H., 1966. Inselbergs and nickpoints in southwestern Arizona. *Z. Geomorph.* 10: 217-225.

Raupach, M., 1957. Investigation into the nature of soil pH. *C.S.I.R.O. Soil Publ.* 9.

Raynal, R. and Nonn, H., 1968. Glacis étagés et formations quaternaires de Galice orientale et de Léon: quelques observations et données nouvelles. *Rev. Géomorph. Dynam.* 18: 97-117.

Read, H.H., 1957. *The Granite Controversy*. Murby, London.

Reid, C., Barrow, G. and Dewey, H., 1910. Geology of the country around Padstow and Camelford. *Geol. Surv. Mem. Explan. Sheets,* pp. 335-336.

Reid, C., 1913. The Tertiary, pp. 102-117. In: W.A.E. Ussher, The Geology of the country around Newton Abbott. *Mem. Geol. Surv. Expl. Sheet* 339.

Reid, C., *et al.,* 1912. The geology of Dartmoor. *Mem. Geol. Surv. Explan. Sheet* 338.

Reusch, H.H., 1883. Note sur la géologie de la Corse. *Paris Soc. Géol. Bull.* 11: 53-67.

Richey, J.E., 1964. Granite. *Mine Quarry Engng.* 30: 20-25, 57-61, 121-127, 167-175.

Richter, G.D. and Kamanine, L.G., 1956. Caractéristique comparative morphologique des boucliers de la partie Européenne de l'URSS, pp. 82-92. In: *Essais de Géographie.* Acad. Sci. URSS, Moscow.

Ritchot, G., 1975. *Essais de Géomorphologie Structurale.* Presses Univ. Laval, Quebec.

Roberts, N.K. and Kallend, P.W., 1978. Selective salt efflorescence as the result of ion exchange on convict-made brickwork at Port Arthur. *J. Austr. Cer. Soc.* 12: 5-7.

Rognon, P., 1967. *Le Massif de l'Atakor et ses Bordures (Sahara Central). Étude Géomorphologique.* C.N.R.S., Paris.

Romanes, J., 1912. Geology of a part of Costa Rica. *Quart. J. geol. Soc. London,* 68: 133-136.

Rondeau, A., 1958. Les 'boules' du granit. *Z. Geomorph.* 2: 211-229.

Rondeau, A., 1961. *Recherches géomorphologiques en Corse.* Colin, Paris.

Rowan, I.S., 1968. Regional gravity survey - Kimba and Elliston 1: 250,000 map areas. *Q. Geol. Notes, Geol. Surv. S. Aust.* 25: 3-4.

Russell, R.J., 1932. Landforms of San Gorgiono Pass, southern California. *Univ. Calif. (Berkeley) Publ. Geogr.* 6: 37-44.

Ruxton, B.P., 1958. Weathering and subsurface erosion in granite at the piedmont angle, Balos, Sudan. *Geol. Mag.* 45: 353-377.

Ruxton, B.P. and Berry, L.R., 1957. Weathering of granite and associated erosional features in Hong Kong. *Geol. Soc. Amer. Bull.* 68: 1263-1292.

Ruxton, B.P. and Berry, L.R., 1961. Notes on faceted slopes, rock fans and domes on granite in the east-central Sudan. *Amer. J. Sci.* 259: 194-206.

Sandford, K.S., 1933. Geology and geomorphology of the southern Libyan Desert, pp. 213-218. In: Bagnold R.A., 1933, A further journey through the Libyan Desert. *Geogr. J.* 82: 211-235.

Sapper, K., 1935. *Geomorphologie der feuchten Tropen.* Teubner, Leipzig.

Saussure, H.B. de, 1796. *Voyage dans les Alpes.* Fauche, Neuchatel.

Sbar, M.L. and Sykes, G.R., 1973. Contemporary compressive stress and seismicity in eastern North America: an example of intraplate tectonics. *Geol. Soc. Amer. Bull.* 84: 1861-1881.

Scheffer, F., Meyer, B. and Kalk, E., 1963. Biologische Ursachen der Wüstenlackbildung. *Z. Geomorph.* 7: 112-119.

Scherber, R., 1932. Erosionswirkungen an der toskanischen Felskuste. *Natur u. Museum.* 62: 231-234.

Schmidt-Thomé, P., 1943. Karrenbildung in kristallinen Gesteinen. *Z. Deutsch. Geol. Gesell.* 95: 53-56.

Scholtz, D.L., 1947. On the Younger Pre-Cambrian granite plutons of the Cape Province. *Trans. geol. Soc. S. Afr.* 49: xxv-lxxxll.

Schrepfer, H., 1933. Inselberge in Lappland und Neufundland. *Geol. Rdsch.* 24: 137-143.

Schumm, S.A., 1963. Disparity between present rates of denudation and orogeny. *U.S. geol. Surv. Prof. Paper.* 454.

Scott, G.R., 1963. Quaternary geology and geomorphic history of the Kassler Quadrangle, Colorado. *U.S. geol. Surv. Prof. Paper* 421-A.

Scott, G.D., 1967. Studies of the lichen symbiosis 3. The water relations of lichens on granite kopjes in central Africa. *The Lichenologist* 3: 368-385.

Scott, W.B., 1897. *An Introduction to Geology.* MacMillan, London.

Scrivenor, J.B., 1931. *Geology of Malaya.* MacMillan, London.

Selby, M.J., 1977. Bornhardts of the Namib Desert. *Z. Geomorph.* 21: 1-13.

Shaler, N.S., 1869. Notes on the concentric structure of granite rocks. *Proc. Boston Soc. Nat. Hist.* 12: 289-293.

Shaler, N.S., 1887-8. The geology of Cape Ann, Massachusetts. *U.S. geol. Surv. 9th Annual Report,* pp. 537-611.

Shannon, C.H.C., 1975. Pseudokarst caves in duricrust/granite terrain, Banana Range, central Queensland, pp.20-24. In *Proc. 10th Bienn. Conf., Austr. Speleo. Fed.,* Brisbane, Dec. 1976.

Sharp, R.P., 1954. Physiographic features of faulting in southern California. In: R.H. Jahn (Editor) *Geology of Southern California. California Div. Mines Bull.* 170: 21-22.

Shaw, P., 1980. Cave development on a granite inselberg, South Rupununi Savannas. *Z. Geomorph.* 24: 68-76.

Simonen, A. and Mikkola, A., 1980. Finland, pp. 51-126. In: *Geology of the European Countries. (Denmark, Finland, Iceland, Norway, Sweden).* Dunod, Paris.

Skertchley, S.B.J., 1893. *Our Island. A Naturalist's Description of Hong Kong.* Kelly and Walsh, Hong Kong.

Smith, B.J., 1978. The origin and geomorphic implications of cliff-foot recesses and tafoni on limestone hamadas in the north-west Sahara. *Z. Geomorph.* 22: 21-43.

Smith, J.M.B. and Lawry, J.R., 1968. Further exploration and observations on Mount Kinabalu East. *Malay Nat.* 22: 39-40.

Smith, L.L., 1941. Weather pits in granite of the southern Piedmont. *J. Geomorph.* 4: 117-127.

Soen Oen Ing, 1965. Sheeting and exfoliation in the granites of Sermersôq, South Greenland. *Med. Grønland* 179 (6).

Sosman, R.B., 1916. Types of prismatic structure in igneous rocks. *J. Geol.* 24: 215-234.

Sparrow, G.W.A., 1961. *The Physiography of a Transition zone. (A contribution to the geomorphology of Southern New England).* Unpubl. M.Sc. Thesis Dept. Geog. Univ. New England, Armidale.

Sprigg, R.C., Wilson, B. and Coats, R.P., 1959. *Alberga.* Map Sheet G 53-9, 1:253,440. Geol. Surv. S. Aust.

Stapledon, D.H., 1961. *Geol. studies for the planning and construction of Tumut 2 underground power station.* Unpubl. M.Sc. thesis, Univ. Adelaide.

Stapledon, D.H., 1966. Geological investigations at the site for Kangaroo Creek Dam, South Australia. *Inst. Eng. Aust., Site Investigation Symposium, Paper 2140, Sydney, Sept. 1966.*

Steers, J.A., 1932. *The Unstable Earth: some recent views in geomorphology.* Methuen, London.

Stockbridge, H.E., 1888. *Rocks and Soils.* Wiley, New York.

Streckeisen, A.L., 1967. Classification and nomenclature of igneous rocks. *Neues Jahrb. Miner. Abh.* 107: 144-240.

Streckeisen, A., 1974. Classification and nomenclature of plutonic rocks. *Geol. Rdsch.* 63: 773-786.

Sugden, D.E. and Watts, S.H., 1977. Tors, felsenmeer and glaciation on Northern Cumberland Peninsula, Baffin Island. *Can. J. Earth Sci.* 14: 2817-2823.

Sutton, D.J. and White, R.E., 1968. The seismicity of South Australia. *J. geol. Soc. Aust.* 15: 25-32.

Sweeting, M.M., 1972. *Karst Landforms.* Macmillan, London.

Syers, J.K. and Iskandar, I.K., 1973. Pedogenetic significance of lichens, pp. 225-248. In: V. Ahmadjian and M.E. Hale (Editors) *The Lichens.* Academic Press, New York.

Taber, S., 1916. The growth of crystals under external pressure. *Amer. J. Sci.* 41: 532-556.

Talobré, J., 1957. *La Mécanique des Roches Appliquée aux Travaux Publiques.* Dunod, Paris.

Tanner, V., 1944. *Outlines of the Geography, Life and Customs of Newfoundland-Labrador.* Acta Geogr. (18.1), Helsinki.

Tarr, R.S., 1891. The phenomena of rifting in granite. *Amer. J. Sci.* 41: 267-272.

Tarr, W.A., 1915. A study of some heating tests, and the light they throw on the cause of the disaggregation of granite. *Econ. Geol.* 10: 348-367.

Terzaghi, K., 1950. Mechanism of landslides. *Geol. Soc. Amer. (Berkey Vol.),* pp. 83-123.

Thiele, E.L. and Wilson, R.C., 1915. Portuguese East Africa between the Zambesi River and the Sabi River: a consideration of the relation of its tectonic and physiographic features. *Geogrl. J.* 45: 16-45.

Thomas, M.F., 1965. Some aspects of the geomorphology of domes and tors in Nigeria. *Z. Geomorph.* 9: 63-81.

Thomas, M.F., 1966. Some geomorphological implications of deep weathering patterns in crystalline rocks in Nigeria. *Trans. Inst. Brit. geogr.* 40: 173-191.

Thomas, M.F., 1967. A bornhardt dome on the plains near Oyo, Western Nigeria. *Z. Geomorph.* 11: 239-261.

Thomas, M.F., 1974a. *Tropical Geomorphology*. Macmillan, London.

Thomas, M.F., 1974b. Granite landforms: a review of some recurrent problems of interpretation. *Inst. Brit. geogr. Spec. Publ.* 7: 13-37.

Thomson, J., 1863. On the disintegration of stones exposed in buildings and otherwise exposed to atmospheric influences. *Brit. Assoc. Adv. Sci. Rept. 32nd Meeting Notices and Abstr.* Murray, London.

Thorbecke, F., 1915. Geographische Arbeiten in Tikar und Wute auf eine Forschungsreise durch Mittel - Kamerun (1911-13). *Verh. 19 deutsch. Geographentag, Berlin, 1915*, pp. 30-41.

Thorbecke, F., 1927. Der Formenschatz im periodisch trockenen Tropenklima mit überwiegender Regenzeit. *Düsseldorfer Geogr. Vort. Erört.* 3: 10-17.

Thorp, M.B., 1969. Some aspects of the geomorphology of the Air Mountains, southern Sahara. *Trans. Inst. Brit. geogr.* 47: 25-46.

Thury, H. de, *et al.*, 1828. Sur le procédé proposé par M. Brard, paur reconnaître, immediatement les pierres que ne peuvent pas résister le gelée. *Ann. Chim. et Phys.* 38: 160-192.

Toit, A.L. du, 1939. *Geology of South Africa*. Oliver and Boyd, Edinburgh.

Trendall, A.F., 1962. The formation of 'apparent peneplains' by a process of combined lateritisation and surface wash. *Z. Geomorph.* 6: 183-197.

Tricart, J., 1957. Mise en point: l'évolution des versants. *L'Inf. Géogr.* 21: 108-115.

Tricart, J., 1962. Observations de morphologie littorale à Mamba Point (Monrovia, Liberia). *Erdkunde*, 16: 49-57.

Tschang Hsi-Lin, 1961. The pseudokarren and exfoliation forms of granite on Pulau Ubin, Singapore. *Z. Geomorph.* 5: 302-312.

Tschang Hsi-Lin, 1962. Some geomorphological observations in the region of Tampin, southern Malaya. *Z. Geomorph.* 6: 253-259.

Tuan, Yi-Fu., 1959. Pediments in southeastern Arizona. *Univ. Calif. (Berkeley) Publ. Geogr.* 13: 1-164.

Turner, F.J. and Verhoogen, J., 1960. *Igneous and Metamorphic Petrology*. McGraw-Hill, New York.

Turner, H.W., 1894. Rocks of the Sierra Nevada. *U.S. geol. Surv. 14th Ann. Rept. Pt. 2*, pp.435-495.

Twidale, C.R., 1955. Pediments at Naraku, north-west Queensland. *Austr. geogr.* 6: 40-42.

Twidale, C.R., 1956. Chronology of denudation in northwest Queensland. *Geol. Soc. Amer. Bull.* 67: 667-687.

Twidale, C.R., 1959. Evolution des versants dans la partie centrale du Labrador-Nouveau, Québec. *Ann. Géogr.* 68: 54-70.

Twidale, C.R., 1960. Some problems of slope development. *J. geol. Soc. Aust.* 6: 131-147.

Twidale, C.R., 1962. Steepened margins of inselbergs from north-western Eyre Peninsula, South Australia. *Z. Geomorph.* 6: 51-69.

Twidale, C.R., 1964. Contribution to the general theory of domed inselbergs. Conclusions derived from observations in South Australia. *Trans. & Papers Inst. Brit. geogr.* 34: 94-113.

Twidale, C.R., 1967. Origin of the piedmont angle, as evidenced in South Australia. *J. Geol.* 75: 393-411.

Twidale, C.R., 1968a. *Geomorphology, with Special Reference to Australia*. Nelson, Melbourne.

Twidale, C.R., 1968b. Origin of Wave Rock, Hyden, Western Australia. *Trans. Royal Soc. S. Aust.* 92: 115-123.

Twidale, C.R., 1971a. *Structural Landforms*. Aust. Natl. Univ. Press, Canberra.

Twidale, C.R., 1971b. Pearson Island Expedition 1969 - 5. Geomorphology. *Trans. Royal Soc. S. Aust.* 95: 123-130.

Twidale, C.R., 1972a. Evolution of sand dunes in the Simpson Desert, central Australia. *Trans. Inst. Brit. geogr.* 56: 77-109.

Twidale, C.R., 1972b. Flared slopes, scarp-foot weathering and the piedmont angle; comparisons between Australia, southern Africa and the western United States. *S. Afr. geogr.* 4: 45-52.

Twidale, C.R., 1973. On the origin of sheet jointing. *Rock Mechanics* 5: 163-187.

Twidale, C.R., 1976a. On the survival of palaeoforms. *Amer. J. Sci.* 276: 77-94.

Twidale, C.R., 1976b. *Analysis of Landforms*. Wiley, Sydney.

Twidale, C.R., 1978a. Granite platforms and the pediment problem, pp. 288-304. In: J.L. Davies and M.A.J. Williams (Editors) *Landform Evolution in Australasia*. Austr. Natl. Univ. Press, Canberra.

Twidale, C.R., 1978b. On the origin of Ayers Rock, central Australia. *Z. Geomorph. Suppl.-Band.* 31: 177-206.

Twidale, C.R., 1978c. On the origin of pediments in different structural settings. *Amer. J. Sci.* 278: 1138-1176.

Twidale, C.R., 1979. The character and interpretation of some pediment mantles. *Sediment. Geol.* 22: 1-20.

Twidale, C.R., 1980a. Origin of minor sandstone landforms. *Erdkunde* 34: 219-224.

Twidale, C.R., 1980b. The origin of bornhardts. *J. geol. Soc. Aust.* 27: 195-208.

Twidale, C.R., 1980c. The Devil's Marbles, central Australia. *Trans. Royal Soc. S. Aust.* 104: 41-49.

Twidale, C.R., 1980d. Landforms, pp. 13-41. In: D.W.P. Corbett (Editor) *A Field Guide to the Flinders Ranges*. Rigby, Adelaide.

Twidale, C.R., 1981a. Granite inselbergs: domed, block-strewn and castellated. *Geogrl. J.* 147: 54-71.

Twidale, C.R., 1981b. Inselbergs - exhumed and exposed. *Z. Geomorph.* 25: 219-221.

Twidale, C.R., 1981c. Origins and environments of pediments. *J. geol. Soc. Aust.* 28: 423-434.

Twidale, C.R., 1982. Les inselbergs en gradin: l'exemple de l'Australie. *Ann. Géog.*

Twidale, C.R., and Bourne, J.A., 1975a. The subsurface initiation of some minor granite landforms. *J. geol. Soc. Aust.* 22: 477-484.

Twidale, C.R. and Bourne, J.A., 1975b. Episodic exposure of inselbergs. *Geol. Soc. Amer. Bull.* 86: 1473-1481.

Twidale, C.R. and Bourne, J.A., 1975c. Geomorphological evolution of part of the eastern Mount Lofty Ranges, South Australia. *Trans. Royal Soc. S. Aust.* 99: 197-209.

Twidale, C.R. and Bourne, J.A., 1976a. Origin and significance of pitting on granite rocks. *Z. Geomorph.* 20: 405-416.

Twidale, C.R. and Bourne, J.A., 1976b. The shaping and interpretation of large residual granite boulders. *J. geol. Soc. Aust.* 23: 371-381.

Twidale, C.R. and Bourne, J.A., 1977. Rock doughnuts. *Rev. Géomorph. Dynam.* 26: 15-28.

Twidale, C.R. and Bourne, J.A., 1978a. Bornhardts developed in sedimentary rocks, central Australia. *S. Afr. geogr.* 60: 34-50.

Twidale, C.R. and Bourne, J.A., 1978b. A note on cylindrical gnammas or weather pits. *Rev. Géomorph. Dynam.* 26: 135-137.

Twidale, C.R., Bourne, J.A. and Smith, D.M., 1974. Reinforcement and stabilisation mechanisms in landform development. *Rev. Géomorph. Dynam.* 23: 115-125.

Twidale, C.R., Bourne, J.A. and Smith, D.M., 1976. Age and origin of palaeosurfaces on Eyre Peninsula and in the southern Gawler Ranges, South Australia. *Z. Geomorph.* 20: 28-55.

Twidale, C.R., Bourne, J.A. and Twidale, N., 1977. Shore platforms and sealevel changes in the Gulfs region of South Australia. *Trans. Royal Soc. S. Aust.* 101: 63-74.

Twidale, C.R. and Corbin, E.M., 1963. Gnammas. *Rev. Géomorph. Dynam.* 14: 1-20.
Twidale, C.R. and Harris, W.K., 1977. On the age of Ayers Rock and the Olgas, central Australia. *Trans. Royal Soc. S. Aust.* 101: 45-50.
Twidale, C.R. and Smith, D.L., 1971. A "perfect desert" transformed. The agricultural development of northwestern Eyre Peninsula, South Australia. *Austr. geogr.* 11: 437-454.
Twidale, C.R. and Sved, G., 1978. Minor granite landforms associated with the release of compressive stress. *Aust. geogrl. Studies* 16: 161-174.
Tyrrell, G.W., 1928. Geology of Arran. *Mem. Geol. Surv. Scot.*
Ule, W., 1925. *Quer durch Sudamerika.* Quitzow, Lubeck.
Urbani, F., 1977. Novedades sobre estudios realizados en las formas carsicas y pseudocarsicas des Escudo de Guayana, Octobre 1977. *Bol. Soc. Venezolana Espel.* 8 (16): 175-197.
Valverde, D., 1968. O Sitio da Cidade, pp. 3-14. In: Curso de Geografia da Guanabara. *Inst. Bras. Geogr.* Rio de Janeiro.
Vogt, J.H.L., 1875. Sheets of granite and syenite in their relation to the present surface. *Geol. Foren Forh.* 56.
Wagner, P.A., 1913. Negative spheroidal weathering and jointing in a granite of southern Rhodesia. *Trans. geol. Soc. S. Africa,* 15: 155-164.
Wahrhaftig, C., 1965. Stepped topography of the southern Sierra Nevada, California. *Geol. Soc. Amer. Bull.* 76: 1165-1190.
Wahrhaftig, C., 1970a. Geologic maps of the Fairbanks A-2-3-4-5 Quadrangles, Alaska. *U.S. geol. Surv. Maps,* CQ-808-811.
Wahrhaftig, C., 1970b. Geologic maps of the Healy D-2-3-4-5 Quadrangles, Alaska. *U.S. geol. Surv. Maps* CQ-804-807.
Wall, J.D.R. and Wilford, G.E., 1966. A comparison of small-scale solution features on micro-granite and limestone in west Sarawak, Malaysia. *Z. Geomorph.* 10: 462-468.
Walther, J., 1900. *Das Geselz der Wustenbildung in Gegenwart und Vorzert.* Reimer, Berlin.
Warth, H., 1895. The quarrying of granite in India. *Nature,(London)* (51): 272.
Watanabe, T., Yamasaki, M., Kojima, G., Nagaoka, S. and Hirayama, K., 1954. Geological Study of damage caused by atomic bombs in Hiroshima and Nagasaki. *Jap. J. Geol. & Geogr.* 24: 161-170.
Waters, R.S., 1954. Pseudo-bedding in the Dartmoor Granite. *Trans. geol. Soc. Cornwall,* 18: 456-462.
Waters, R.S., 1957. Differential weathering and erosion on oldlands. *Geogrl. J.* 123: 503-509.
Waters, R.S., 1964. The Pleistocene legacy to the geomorphology of Dartmoor, pp. 73-96. In: I.G. Simmons (Editor) *Dartmoor Essays.* Devon. Assoc. Adv. Sci. Lit. & Art.
Watts, S.H., 1979. Some observations on rock weathering, Cumberland Peninsula, Baffin Island. *Can. J. Earth Sci.* 16: 977-983.
Wayland, E.J., 1934. Peneplains and some erosional landforms. *Geol. Surv. Uganda, Ann. Rept. Bull.* 1: 77-79.
Wayland, E.J., 1953. More about the Kalahari. *Geogrl. J.* 119: 49-56.
Weed, W.H. and Pirsson, L.V., 1898. Geology and mineral resources of the Judith Mountains of Montana. *U.S. geol. Surv. 18th Ann. Rept. (Pt. 3 Economic Geology):* 445-616.
Wellman, H.W. and Wilson, A.T., 1965. Salt weathering: a neglected geological erosive agent in coastal and arid environments. *Nature* 205 (4976): 1097-1098.
Whitaker, C.R., 1973. *Pediments, a bibliography.* Geo Abstracts, Norwich.
Whitaker, C.R., 1974. Split boulders. *Aust. geogr.* 12: 562-563.
Whitaker, C.R., 1978. Pediment form and evolution in the East Kimberleys, pp pp. 305-330. In: J.L. Davies and M.A.J. Williams (Editors) *Landform Evolution in Australasia.* Austr. Natl. Univ. Press., Canberra.
Whitaker, C.R., 1979. The use of the term "pediment" and related terminology. *Z. Geomorph.* 23: 427-439.
White, S.E., 1973. Is frost action really only hydration shattering? A review. *Arctic & Alp. Res.* 8: 1-6.

White, W.A., 1945. Origin of granite domes in the south-eastern Piedmont. *J. Geol.* 53: 276-282.

White, W.B,, Jefferson, G.L. and Haman, J.F., 1966. Quartzite karst in south-eastern Venezuela. *Inter. J. Speleology* 2: 309-314.

White, W.S., 1946. Rock bursts in the granite quarried at Barre, Vermont. *U.S. geol. Surv. Circ.* 13.

Whitney, J.D., 1865. *Geology of California,* Vol. I. Calif. State Dept. Mines, San Francisco.

Wilhelmy, H., 1958. *Klimamorphologie der Massengesteine.* Westermann, Brunswick.

Wilhelmy, H., 1964. Cavernous rock surfaces (tafoni) in semiarid and arid climates. *Pak. geogrl. Rev.* 19: 9-13.

Wilhelmy, H., 1974. *Klima-Geomorphologie in Stichworten.* Hirt, Kiel.

Williams, G., 1936. The geomorphology of Stewart Island, New Zealand, *Geogrl. J.* 87: 328-337.

Williams, G.E., 1968. Torridonian weathering, and its bearing on Torridonian palaeoclimate and source. *Scot. J. Geol.* 4: 164-184.

Williams, G.E., 1969. Characteristics and origin of a Precambrian pediment. *J. Geol.* 77: 183-207.

Williams, I.S., Compston, W., Chappell, B.W. and Shirahase, T., 1975. Rubidium-strontium age determinations on micas from a geologically controlled, composite batholith. *J. geol. Soc. Aust.* 22: 497-505.

Willis, B., 1934. Inselbergs. *Assoc. Amer. geogr. Ann.* 24: 123-129.

Willis, B., 1936. East African plateaus and rift valleys. In: *Studies in Comparative Seismology.* Carnegie Inst. Washington D.C. Publ. 470.

Winkler, E.M., 1965. Weathering rates as exemplified by Cleopatra's Needle in New York. *J. geol. Educ.* 13: 50-52.

Winkler, E.M., 1975. *Stone: Properties, Durability in Man's Environment.* Springer, New York.

Winkler, E.M., 1979. Role of salts in development of granitic tafoni, South Australia: discussion. *J. Geol.* 87: 119-120.

Wojcik, Z., 1961a. Caves in granite in the Tatra Mountains. *Actes 3 ère Congr. Inter. Spéléol. (Salzburg)* A: 43-44.

Wojcik, Z., 1961b. Karst phenomena and caves in the Karkonosze granites. *Die Höhle* 12: 76.

Wolff, R.G., 1967. Weathering of Woodstock granite, near Baltimore, Maryland. *Amer. J. Sci.* 265: 106-117.

Wolters, R., 1969. Zur Ursache der Entstehung oberflächenparalleler Klüfe. *Rock Mechanics* 1: 53-70.

Woodard, G.D., 1955. The stratigraphic succession in the vicinity of Mount Babbage Station, South Australia. *Trans. Royal Soc. S. Aust.* 78: 8-17.

Wopfner, H., Callen, R.A. and Harris, W.K., 1974. The Lower Tertiary Eyre Formation of the southweastern Great Artesian Basin. *J. geol. Soc. Aust.* 21: 17-51.

Wopfner, H. and Twidale, C.R., 1967. Geomorphological history of the Lake Eyre Basin, pp. 118-143. In: J.N. Jennings and J.A. Mabbutt (Editors) *Landform Studies from Australia and New Guinea.* Austr. Natl. Univ. Press, Canberra.

Worden, J.M. and Compston, W., 1973. A Rb-Sr isotopic study of weathering in the Mertondale granite, Western Australia. *Geochim. & Cosmochim. Acta* 37: 2567-2576.

Worth, R.H., 1953. *Dartmoor.* G.M. Spooner and R.S. Russell (Editors). David and Charles, Newton Abbott.

INDEX

acanalduras, 237
acid weathering, 65, 66
Ackermann, E., 96, 107
Ackermann, H.D., 193
acuminate boulder, 247, 254
Adamaua, West Africa, 125
adamellite, 39
Adams, J.A.S., 317
Adams, J.B., 298
Adirondack Mts, N.Y. State, 64
Adnan, B.A.N., 298
African Surface, 128, 164
Agassiz, L., 23, 71, 137, 181
age of inselbergs, 146-147
Agulhas Negras, Brazil, 178
Ahnert, F., 181
Air Mts, Sahara, 125
Akagi, Y., 192
Alabama Hills, California, 1, 245
Alexander, F.E.S., 268, 271
algae, 217
Algeria, 13
Alice Springs, N.T., 16
    nubbins, 164
alignment of residuals hills, 140
alkaline weathering, 65, 275
Allen, J.R.L., 234
all-slopes, 5, 17, 18, 22, 177
    and equilibrium conditions, 181-183
    distribution, 177-179
    in cold lands, 179-181
    origins, 179 *et seq.*
alluvial plain, 205
alveolar weathering, 286, 287, 288
alveoles, 286, 287, 288
Amaral, I.do, 165
Ambrose, J.W., 187
Amstutz, G.C., 119
ancient granite forms, 146-147, 336
Anderson, A.L., 17, 68, 96, 215, 298
Andes Mts, S. America, 17, 177
Andrée, D.V. d', 119
Andrée, K., 295
Angola, 12, 147, 165
Anhaeusser, C.R., 139
annular depression, 258, 264
Antarctica, 289
antiquity of granite forms, 146, 147
anvil rock, 247 255
aplite sill, 85
araceenhorst, 217
arched slab, 319
arcuate joints (see also sheeting
    joints), 43

arête, 179
Armchair, The, S.A., 182
armchair-shaped hollow, 214, 218
Arnold, J., 280
Arran, Scotland, 98, 150
Ashburton R., W.A., 238
Aswan, Egypt, 39
A-tent, 317 *et seq.*
    expansion involved, 321
    in rock other than granite, 318
atomic explosion
    and granite weathering, 61
Ava ring intrusion, Finland, 38
Ayers Rock, N.T., 124, 195, 198, 199,
    247, 259
azonality, of granite forms, 24, 336

Badgley, P., 29
Bagnold, R.A., 207
Bahia Guanabara, Brazil, 7, 136
Bakker, J.P., 146, 217, 268, 269, 275,
    294
Bain, A.D.N., 60, 72, 137, 209
balanced rock, 95
Balanced Rock, Texas, 104
balancing rock, 95, 103
Balancing Rocks, Zimbabwe, 103
Bald Rock, Vic., 27
Balk, R., 43, 119, 155
Balladonia, W.A., 261
Baltic Shield, 30
Bankung, 49
Barbeau, J., 113
Barbier, R., 71, 125, 155, 336
Bardin, V.I., 295
Barre, Vt., 51, 57
barrel-like corestone, 111, 114, 116
Barrère, P., 125, 177
Barton, D.C., 60, 61, 72, 77, 209
Bartrum, J.A., 295
basal depression, 258
basal fretting, 257-258
basal tafoni, 235
Bateman, P.C., 30, 151
bath tub, 215
batholiths, 29
    S.E. Australia, 31
    Brittany, 32
    Armorican Massif, 32
bathylith (see batholith), 29
Bauer, M., 274
Baulig, H., 207, 330
Baumverfallspingen, 217
bauxite, on granite, 59

Beche, H.T. de la, 43, 100, 101, 102, 154
'bedded' granite, 48-49, 222-223
Bellever Tor, Dartmoor, 306, 307
Bendefy, L., 156
Berger, A.R., 29
Bergfussniederungen, 262
Berliat, K., 71
Berridale Batholith, N.S.W, 3
Berry, L.R., 117, 262
Bieniawski, J.T., 135
Bigarella, J., 269
billiard table surface, 258, 261
biotite, weathering of, 67, 68
Bird, J.B., 318
Birot, P., 61, 67, 74, 82, 146, 155, 191, 206, 268, 271, 292
Black Hill, S.A., 26, 68, 70
  welded joints, 83
Blackingstone Rock, Dartmoor, 132, 154, 158, 161
Blackwelder, E., 61, 151, 200, 203, 298, 307, 314
blade-like form, 247, 254
Blank, H.R., 233, 234, 317
blister, 317
Bloom, A.L., 23, 137, 332
Blue Granite, Dartmoor, 81, 82
Blue Tier Batholith, Tasmania, 30
Blyth, F.G.H., 3, 55, 155
Boase, W., 154
Bocquier, G., 263, 265
Bohemian Massif, 16, 73, 152, 220
Boland, J.N., 48, 67
Boone's Cave, N. Carolina, 287
bornhardts, 8, 11, 12, 13, 15, 16, 21, 124 *et seq.*
  and compression, 131, 134-135
  and folds, 131, 134
  and fluvial erosion, 137
  and glaciation, 137
  and multicyclic landscapes, 125, 128, 147-148
  and scarp retreat, 138 *et seq.*
  and sheet structure, 150-158
  as corestones, 124, 125, 131
  as plutonic margins, 128, 131
  azonal forms, 149
  characteristics, 124 *et seq.*
  coastal, 136
  convergent forms, 334
  domical shape, 149-158
  environments of development, 136 *et seq.*
  in non-granitic rocks, 124-125
  in upland settings, 139
  lithological, 125-130
  marginal attack of, 169 *et seq.*
  multicyclic landscapes, 125, 128, 147
  multiphase development, 145-146

bornhardts (cont'd)
  subsurface initiation, 135 *et seq.*
  tectonic, 125
  two stage development, 135 *et seq.*
  upfaulted, 125
  why upstanding, 125 *et seq.*
Bornhardt, W., 7, 8, 124, 136, 137
Booth, B., 34
Borlase, W., 214, 217
botryoidal structure, 299
Bott, M.H.P., 29, 154
boulder, 5, 6, 9, 10, 17 (definition), 22, 23, 68, 89 *et seq.*, 333
  due to disintegration of sheet structure, 119-121
  of decomposition, 108
  of disintegration, 110
  origin, 89 *et seq.*
  perched, 95
  shape, 95
  size of, 89, 95
boulder beach, 2
boulder tafoni, 286
boule, 92
Bourcart, J., 294
Bourne, J.A., 44, 77, 124, 147, 189, 199, 200, 226, 228, 233, 236, 238, 240, 245, 262, 288, 291, 308, 312, 322, 327, 328, 329.
Bowerman's Nose, Dartmoor, 108
bowlder (see also boulder), 97
Boyé, M., 20, 137, 142, 200, 228, 276
Bradley, W.C., 51, 153, 295
Brajnikov, B., 131, 154
Brammall, A., 81, 154
Branner, J.C., 71, 77, 81, 108, 269, 270
Brazil, (Poondana) Rock, S.A., 212
Bremer, H., 124, 276
Bridgman, P.W., 64, 112
Brown, D.A., 136
Browne, W.R., 16
Bruckner, W.D., 294
Brunner, F.K., 152, 157
Brunsden, D., 92, 192
Bryan, K., 200
Buccleuch, R.S.A., 143
Buckley, H.E., 295
Büdel, J., 23, 132, 135
Bulow, K., 237
Bureau Recherches Géol. Minières, 32
buried plain, 189
Bury Hills, Zimbabwe, 246
Buseck, P.R., 67, 70, 116
Bushfeld Lopolith, R.S.A., 37
Bushman Surface, R.S.A., 12, 22, 205
Bushman's Kop, R.S.A., 128
Bussersteine, 96, 107
butte, 182

Caesalpus, 36
Cailleux, A., 288, 295, 296

Caine, N., 173
Cairnsmore of Fleet pluton,
    Scotland, 29, 38
Caldcleugh, A., 281
caldeiros, 217
Calkin, P., 288
Callen, R.A., 184, 278
Cameron, J., 55
Canadian Shield (see Laurentian
    Shield), 29
cannelures, 237
Cape Ann, Mass., 44
Cape Town, R.S.A., 187, 188
Carappee Hill, S.A., 320
Carey, S.W., 45
Carl, J.D., 119
Carlé, W., 270, 274
Carnegie, D.W., 215
case hardening, 58, 297 *et seq.*
Cassia City of Rocks, Idaho, 17
castellated form
    on dome, 158, 161
castle koppie, 14, 15, 124, 158, 160,
    167 *et seq.*
    and structure, 168
    in cold lands, 170, 173 *et seq.*
Castle Peak, B.C., 35
Castle Rock, W.A., 169
Cathedral Rocks, California, 177
cauldron, 215, 217
cave, 280
    and fractures, 281
    and grus, 280
cavitation, 273
Chapman, C.A., 69, 116, 151
Chayes, F., 36
cheesering, 95, 101
cheesewring, 95
chemical weathering, 64-68
    in cold climate, 73
Cheung Chau, Hong Kong, 289
Chinaman's hat, 247, 253
Chilpuddie Hill, S.A., 250, 256
Choffat, P., 291
Choubert, B., 45, 47, 125, 155
Cima Dome, California, 191, 194
Cimatino Dome, California, 191, 194
circular structures in gneiss, 155
Clayton, R.W., 199, 263
cleft (see also slot), 15, 87, 145,
    213
    with flares, 252
cliff floor cave, 200, 286
climate and weathering, 71-73
climates of past, 335
    of Cainozoic, 24
    of Mesozoic, 24
climatic indications (granitic), 23
climatic zonation, 331, 337
clitter, 62
Cloos, E., 45, 155

Cloos, H., 43, 45, 48
Cloudesley-Thompson, J.L., 314
coastal batholith, Peru, 30
coastal bornhardts, 136
coastal forms, granite, 2
Coates, D.F., 156, 318, 327
Cobbing, E.J., 30
Cole, W.F., 294
Colorado Plateau, U.S.A., 153
compayrés, 94, 99
compressive stress, 327, 328
Compston, W., 336
concentric structure, 113
connate salts, 295
convergence, 333
convergent development, 333
Cooke, R.U., 292, 294, 295, 297
Coolie hat, 247, 253
Corbel, J., 138, 206
Corbin, E.M., 215, 219, 227, 238
Corcovado, Brazil, 7, 74, 76
corestone, 2, 5, 70, 91-98, 109-110
    and volume increase, 118, 119
    as primary petrological feature, 89,
        113
    anomalous, 117, 118
    development, 89-93
    in other rock types, 93
    in rhyolite matrix, 113
Corndon, Shropshire, 37
Corrigin, W.A., 22
Corrobinnie Depression, S.A., 140
Corrobinnie fault zone, S.A., 44
Corrobinnie Hill, S.A., 21, 143, 259
Corrobinnie platform, S.A., 194, 195
Corsica, 75
Costin, A.B., 16
cottage loaf, 95, 106
Cotton, C.A., 16, 138, 207, 260, 262
Cowie, J.W., 187
crazy paving, 88
Crestes de Gargantillar, Pyrenees, 180
Crickmay, C.H., 201
Crickmay, G.W., 228
Crowell, J.C., 134
crystal dislocation, 67, 116
Cumberland Stone, Scotland, 258, 259
Cunningham, F.F., 173
curved joint, 113-114
cylindrical basin, 291
cylindrical hollow, 214, 219, 226
Czudek, T., 73, 220

Daadening Hill, W.A., 308
Dahl, E., 73
Dahl, R., 73, 218, 239, 269,
Dahlke, J., 243
Daily, B., 136
Dale, T.N., 40, 48, 49, 50, 155, 318
Daly, R.A., 35
Darling Ranges, W.A., 59, 139, 143, 147

Darnley, A.G., 55
Dartmoor, 3, 18, 74, 81, 104
    sheeting, 153
Darwin, C.R., 60, 106, 335
Davis, S.N., 66
Davis, W.M., 22, 200, 201, 207, 338
Dayman Dome, P.N.G., 155
Dearman, W.R., 155
Debon, F., 84
decay of granite
    Rio, 71
    Malagassy, 71
decomposition bowlder, 108
Dellen, 217
Demek, J., 16, 136, 167, 173, 175,
    269, 271, 286, 288
demi-oranges, 15
Denaeyer, M.E., 295
Denham, D., 156, 328
depressions de piedmont, 262
depth of weathering, 71
Devil's Marbles, N.T., 58, 88, 101,
    106, 158, 162, 168, 174, 183, 301,
    303, 307
Devil's Slide, Lundy Is., England, 52
dew hole, 288
dike (see dyke), 32
Dinosaur, S.A., 251
disequilibrium forms, 24
displaced block, 317 *et seq.*, 322,
    323
    origin, 327 *et seq.*
displaced slab, 317 *et seq.*, 322,
    323
    origin, 327 *et seq.*
distribution of granite forms, 331-333
Djanet, Tassili Mts, 15
dome, 2, 9, 13, 15
    bornhardt, 149-158
    gneiss, 32, 37
Domeland, Sierra Nevada, California,
    6, 253
Dorman, F.H., 278
Dorn, R.I., 299
dos de baleine, 8
dos d'éléphant, 8
doughnut (see rock doughnut)
    induration, 233, 234
    scour, 234
    relief inversion, 234, 235
Dragovich, D.J., 288, 292, 294, 296
Drake, F.E., 217
Dresch, J., 146, 191, 205
drip pool, 229, 232
Druids, and rock basins, 217
dry reactions (in weathering), 64, 65
Duffaut, P., 113
Dumanowski, B., 74, 80, 177, 262
dumb-bell, 247
Dumonte Rock, S.A., 144, 227
Durand Delga, M., 75

Dury, G.H., 202
dwala, 15
dyke, 32
Dzulinski, S.T., 49, 173, 214

earth structure, 26
earth tremors, 328
Eaton, J.P., 30
Ebaka, South Cameroon, 142, 144, 221,
    228, 288
elephant rock, 8
Eggler, D.H., 173
Eggleton, R.A., 67, 70, 116
Eiseley, L., 335
Elders, W.A., 29
Embleton, B.J.J., 136
Emerson, B.K., 156
Emery, K.P., 61, 314
Enchanted Rock, Texas, 123, 233, 319
Enchanted Rock Cave, Texas, 281, 282
Encounter Bay, S.A., 7, 74
Engel, C.G., 298
environments of bornhardt development,
    136 *et seq.*
equifinality, 333
erratic (glacial), 23, 137
Eskola, P., 37, 155
Esperance Bay, W.A., 8
Estes Park, Colorado, 159
etch form (see also etch surface)
etch plain, 187 *et seq.*
    evolution, 188
etch surface, 59, 184, 336
    W.A., 187-189
    Mount Lofty Ranges, S.A., 189
    Labrador Peninsula, 187
Et-then Series, 4
Evans, I.S., 292, 294
Everard Range, S.A., 6, 10, 16, 77, 124
exfoliation (see also sheet structure)
    48-49
exhumed inselberg, 146
exhumed plain, 187, 188
exsudation, 295
Eyre Peninsula, S.A.
    age of inselbergs, 147

faceted slope, 184
Fairbridge, R.W., 147
Falconer, J.D., 16, 135, 137, 148-149
Farmin, R., 112, 313
faults
    and rim, 86
fault scarp, 1
fault-line scarp, 4
fault-line valley, 2
faulted granite, 1, 2, 3
Feininger, T., 280
feldspar, weathering of, 67, 74
Felschüssel, 217
Fernlinge, 138, 139

fine-grained rocks, resistant, 80
Finkl, C.W., 147
Finlayson, B., 280
Fisher, O., 181
fitted block, 2
fitted boulder, 2
fixation, 65
flaggy structure, 47-51, 222-223
flakes, 70, 111, 286, 292
   and early breakdown of granite, 70
   variations in thickness, 118
flared boulder, 293
flared slope, 17, 19, 24-25, 54, 144
   et seq., 169, 172, 200, 243-257,
   332, 333
   and tafoni, 253, 257, 291
   characteristics, 244
   description, 244
   distribution, 243-244
   in clefts, 252
   multiple, 244, 249, 252
   on non-granitic rocks, 243-244
   origin, 144, 145, 247-257, 332-333
   subsurface development, 144, 145,
      244-257
flat-lying joint, 49
Flinders Ranges, S.A., 6, 17-18, 177,
   181-182, 184-185
floater, 92
fluting, 2, 17-18, 23, 263, 267-279
   and relief inversion, 263, 271
   coastal, 274-275
   congeners in sandstone, 269
   distribution, 269-289
   fluvial, 2, 238
   origin, 270-279
   processes, 270 *et seq.*
   structural contour, 270
fluvial erosion and bornhardts, 137
foliation, 106
   exploited by weathering, 171
Folk, R.L., 194, 259
Ford, T.D., 16
Fort Lamy, Sahara, 183
Fort Trinquet, Mauritania, 120, 191
Foureau, F., 183
fractures in granite, 43-55, 82-88,
   150-158
   and valleys, 15, 127
   and weathering, 82-88
   density, 130 *et seq.*
   in relation to frost shattering, 82
   margins of inselbergs, 139
Francis, W.D., 298
Frederickson, A.F., 66
freeze-thaw, 62, 175
   difficulties, 64
French Guyana, 74, 125
   fractures, 45, 47
Frenzel, G., 294
fretted basal slope, 257-258, 293

fretted block, 2, 238, 241
fretted boulder, 2
Fritsch, P., 20, 142, 200, 228, 276
Frog Island, Singapore, 267
frost-riving, 62, 63, 72
Fry, E.J., 234, 271
Fuge, R., 296
fule, 262

Gage, M., 152
Galli-Olivier, C., 330
Gavea, Brazil, 74, 75
Gebel Harhagit, Egypt, 262-266
Gee, R.D., 30
Geikie, A., 108
Geikie's corestones, 110
Gemencheh, W. Malaysia, 93
gendarme, 179, 180-181
genetic relationships, inselbergs, 175
   et seq.
Geological Society of Australia, 31
Georgina River plain, Q'd., 205
Gerrard, A.J.W., 154
Gerth, H., 177
Gèze, B., 113
Giant Granite, Dartmoor, 81-82
gibber plain, 11
Gifkins, R.C., 153
Gilbert, G.K., 151, 156
Giles, E., 7
Gilluly, J., 146, 156
glacial erratic, 23, 137
glaciated granite outcrops, 2
glaciated valley, 129
glaciation and bornhardt development, 137
gnamma, 215
gneiss dome, 32, 37
gneissic dome, 156-157
Godard, A., 72, 82, 167, 169, 175, 269
Gokomere, Zimbabwe, 161, 163
Goldich, S.A., 73
Goodenough Island, P.N.G., 155
Gosse, W.C., 124
Goudie, A., 292, 294-295
grain, 46, 49
granit pourri, 105
granite
   and faults, 1, 43
   bedded, 48
   bornhardt, 149-158
   coastal, 2
   composition, 36, 40
   compressive strength, 39
   definition, 36
   dome, 2, 9, 13, 15
   flaggy, 48-49, 50-51
   flexibility, 39
   fractures, 43
   glaciated, 2
   gneiss, 32, 37
   hardness, 39

granite (cont'd)
 local names, 39
 occurrences, 26
 origins, 27
 permeability, 39-40
 perviousness, 39-40
 physical characteristics, 39 *et seq.*
 pitted, 62
 porosity, 39-40
 specific gravity, 39
 stable when dry, 59, 94
 unstable when wet, 59, 94
granite forms
 as climatic indicators, 231, 330
 distribution, 331-333
granite pediments, origins, 199 *et seq.*
granite plains, 5, 17, 22, 186 *et seq.*
Granite Rock, Vic., 269, 276
granitic nuclei of continents, 27
Granitrille, 237
Grawe, O.R., 64
Green, P., 66
Greenfield, M.A., 69, 116, 151
Greenland, 2, 147, 177-178, 188
 bornhardt, 137, 139
 palaeosurface, 147, 187, 188
Griffith fracture, 157
Griggs, D.T., 61, 314
Groot Spitzkoppe, Namibia, 11
Groothoekseberg, R.S.A., 11, 17
grooves, 267-279
 and relief inversion, 263, 271
 coastal, 274-275
 congeners in sandstone, 269
 distribution, 269-289
 origin, 270-279
 processes, 270 *et seq.*
 structural control, 270
 subsurface initiation, 275-279
Groves, D.I., 30
growan, 92
grus, 2, 91-95, 97-98, 109-110
gruss - see grus
gutter, 213, 237-242
 description, 237-239
 in glaciated lands, 239
 in river bed, 238, 240
 initiation, 240
 origin, 239 *et seq.*
 rims of, 237
 subsurface initiation, 240-242
 undercut sidewalls, 237

Hack, J.T., 151, 155
Hake, B.E., 209
Hale, M.E., 234
Half Dome, California, 5, 129
half-orange, 9
Halfway, R.S.A., 142, 144, 228
Hambleton-Jones, B.B., 291, 295

Hamersley Surface, W.A., 164, 166, 206
Hamilton district, Labrador, 46
Hamilton, W., 29
Handley, J.R.F. 16, 164
Hankar-Urban, A., 156
hardness, granite, 39
hardway, 46
Harland, W.B., 48, 62, 152
Harris, G.F., 136, 154
Harris, W.K., 278
Harriss, R.C., 316
Hartlinge, 149
Hartt, C.F., 108
Harwood, H.F., 81
Hassenfratz, J-C., 98
Hassenfratz's Site, 109
Haytor, Dartmoor, 19, 82, 170
'heart of the block', 92
heating as means of quarrying, 61
 atomic explosion, 61
 by fire, 60-62
 moisture and weathering, 61
 nuclear explosion, 61
Hedges, J., 227, 281
heiroglyph, 306, 309
Helbig, K., 268, 274
Hellstrom, B., 243
Heltor, Dartmoor, 51
Herbert Falls, Q'd., 9
Herrmann, L.A., 125
Herrmann, O., 52
Hitchcock, C.B., 268
Hills, E.S., 2, 16, 43, 207, 233, 302
Hodgson, J.H., 26, 27
Hoek, E., 135
Högbom, B., 295
Hoggar Mts, Algeria, 158, 160
Holmes, A., 7, 29, 128, 158, 262
honeycomb weathering, 286
hoodoo rock, 238
Hooke, R. Le B., 298
horn, 179
horst, 1
Horwitz, R.C., 206
hourglass boulder, 236, 247
Howard, A.D., 138, 199
Howchin, W., 95
Humboldt, A. von, 298
Hume, W.F., 177, 302, 314
Humps, The, W.A., 147
Hunt, C.B., 35
Hurault, J., 74, 151
Hutton, J., 36, 97, 335
Hutton, J.T., 68, 294, 297
Hyden Rock, W.A., 145
hydration, 64, 116
hydration shattering, 70-72, 119
Hydratationssprengung, 70
hydrolysis, 65, 116
hydrothermal action, 83, 115

imprisoned block, 2
imprisoned boulder, 2
incipient dome, 142-143
induration - and doughnut, 233-234
Ingles, O.G., 152, 313
inherited forms, 330
initial breakdown of granite, 68-71
inselberg, 5, 7, 11, 12, 22, 23, 124
   et seq.
  age, 146-147
  and regional structure, Transvaal,
    139
  and tectonic style, 139
  characteristics, 124
  fracture-controlled margins, 139
  massif, 124, 146-147
  plan shape, 126
  regional setting, 139-140
inselberg de poche, 169
Inselberge (see inselberg), 7
Inselberglandschaft, 7, 12, 186
insolation, 60
intradosal zone, 112
Investigator Group, S.A., 6-7
iron induration, 223
Isaacson, E. de St. Q., 112, 156
Isa Highlands, Q'd., 59, 184
Isherwood, D., 67, 74
Iskander, I.K., 234, 271
island mountain, 6

Jack, R.L., 77
Jaeger, J.C., 328
Jahn, A. 16, 49, 77, 214
Jahns, R.H., 50, 55, 152
Jameson, R., 93, 98, 99
Jeje, L.K., 87
Jennings, J.N., 286, 316, 318, 321,
  327, 328
Jessen, O., 7, 12, 125, 147, 269
Johnson, A.R.M., 281
Johnson, R.J., 307, 310, 311
Johnston, J.H., 294
joints in granite (see also
  fractures), 43 *et seq.*
  and regional structure, 44, 45
  as expression of off-loading, 151
Joly, F., 191
Joly, J., 65, 275, 332
Jones, T.R., 16, 92, 95, 105, 113,
  154, 217
Judith Mts, Montana, 36
Jutson, J.T., 2, 16, 59, 137, 187,
  189, 215, 258, 286, 295, 318

Kallend, P.W., 294
Kamanine, L.G., 45
Kamiesberg, R.S.A., 11, 17, 124, 139,
  157, 192, 208
Karkonosze Mts, Poland, 16, 17, 74,
  177, 281

Kappakoola Depression, S.A., 140
Kappakoola fault zone, S.A., 44
Karibib, Namibia, 80, 86
Karren, 237, 277
Kastning, E.H., 281
Keller, W.D., 66
Kelly, W.C., 295
Kemp, E.M., 136, 278
kernel, 92
Kessler, D.W., 40, 111
Keyes, C.R., 137
Kiersch, G.A., 152
Kieslinger, A., 55, 119, 137, 152, 296
Kilba Hills, Nigeria, 15
King George Sound, W.A., 8
King, L.C., 125, 138, 146, 151, 204, 205,
  206
Kingsmill, T.W., 93, 105, 106
Kinzl, H., 177
Klaer, W., 75, 214, 269, 270, 288, 292,
  295
Kluftkarren, 88, 237, 270, 305
knoll, 124
Kober, L., 207
kociolki, 217
Kokerbin Hill, W.A., 145, 257, 293, 305
Kolar gold field, India, 156
Kotarba, A., 49, 173, 214
Kranck, E.H., 155
Krauskopf, K.B., 209
Krieg, G.W., 10
Kukal, Z., 298
Kulkera Hills, central Australia, 217
Kvelberg, I., 291, 296
Kwaad, F.J.P.M., 294
Kwaterski Rocks, S.A., 87, 219
Kylie Lake, Zimbabwe, 127

Labertouche Cave, Vic., 280
Labrador Peninsula, 187
laccolith, 31, 35, 36
Lagasquie, J.T., 76, 186
Lägerklufte, 49
Lahee, F.H., 338
L. Chad, central Africa, 113, 116
L. Havasu, Arizona, 132
L. Ladoga, Finland, 37
L. Tahoe, California, 91
Lamego, A.R., 52, 74, 125, 131, 134, 155
lamellar structure, 48
lamination, 48
Land's End, England, 7
Lane, E.W., 17
Langford-Smith, T., 202
lapiaz, 237
lapiés, 237
Lapparent, A. de, 291
Lapworth, C., 37
Larsen, E.S., 68, 69, 70, 109
lateral wedge, 327

laterite - on granite, 59, 93
  capped plateaux, 59
lateritised surface, W.A., 147
Laurentian Shield, 29
Laurie, A.P., 294
Lautansach, H., 82
Lawry, J.R., 318
Lawson, A.C., 191, 199, 200
leaching, 65
Le Conte, J.N., 131
Lee, W.T., 207
Leeman, E.F., 112
Leeukop, R.S.A., 80, 142, 143
Lehmann, H., 286
Leigh, C.H., 51, 173, 175
Lei Yue Mun, Hong Kong, 283
Leonard, R.J., 307, 308, 309, 310, 312
Lester, J.G., 125
'levee-like' rim, 237
Leviathan, The, Vic., 89, 90
Lewis, W.V., 94, 152
Libya, 254
Liesegang ring, 119
lift, 47
Lightburn Rocks, S.A., 194, 198, 320
Linton, D.L., 16, 92, 95, 96, 97, 119,
  135, 146, 152, 173, 257
Liskeard, Cornwall, 101
Lister, L.A., 88, 237
Little Shuteye Pass, California, 53
Little Wudinna Hill, S.A., 122
Ljungner, E., 49
Lobek, A.K., 207
Logan, J.R., 23, 91, 102, 103, 110,
  227, 267, 270, 275, 276, 277
Logan, R.F., 295
loganstone, 95
logging stone, 95, 102
Lone Pine, California, 1
lopolith, 31, 37
Loudermilk, J.D., 298
Loughnan, F.C., 64
Louis, H., 303
Ludbrook, N.H., 278
Lundy Is., England, 8, 24

Mabbutt, J.A., 16, 59, 189, 200, 206,
  262
MacCulloch, J., 16, 43, 91, 99, 100,
  214, 217, 218, 227
MacDonald Fault Scarp, N.W.T., 4
MacGregor, A.M., 155
Mackin, J.H., 190, 193
Maclaren, M., 215
MacMahon, C.A., 150
Mallos de Riglos, Spain, 125
mamelon, 182
mamillation, 286, 290
mantle-controlled planation, 200
mantled pediment, 4, 20, 21, 190 et
  seq.

distribution, 191, 192
marginal attack, of bornhardts, 169 et
  seq.
marine salt, 295
marmitas, 217
Marti Boni, C., 83
Martini, I.P., 281
Martonne, E. de, 23, 92, 94, 109
Mason, B., 64, 65, 275, 332
massif (granite), 6
Massif Central, France, 76, 98, 303
matopos, 9
Matopos, Zimbabwe, 131
Matthes, F.E., 48, 151, 209, 214
Matschinski, M., 296
McGee, W.J., 192, 193, 194
Meekatharra Surface, W.A., 23
  ultiplain, 205, 206
meias laranjas, 15
menhir, 258, 260
Mennell, F.P., 131, 132, 149
Mensching, H., 192
meringue surface, 223, 225
Merrill, G.P., 150
Merrivale, Dartmoor, 51
Mertondale granite, W.A., 336
mesa, with Fe rich capping, 58
Metallagenium, 299
Meunier, A., 64
Michel Mainguet, M., 124, 281
microfissures, 82, 114
microfractures, 66, 67
Mikkola, A., 30, 38
Miller, W.J., 291
mineral banding, 113, 115
Mistor Pan, Dartmoor, 220
moat, 231, 262
moisture
  attack, 24, 59, 64, 88, 334
  and basins, 219, 227, 228
Mojave Desert, California, 265
monadnock de dureté, 149
monadnocks de position, 138
monadnock de résistance, 149
monkstone, 96
monzonite, 39
morros, 9, 154, 182
Mortensen, H., 294
Moss, A.J., 61, 66, 67, 71
Mount Lofty Ranges, S.A., 107, 158
  etch surface, 189
Mt Bresnahan, W.A., 124
Mt Bundey, N.T., 16, 161, 164, 166
Mt Hall, S.A., 283
Mt Hillers Stock, Utah, 35
Mt Kobe, Mozambique, 158
Mt Manypeaks, W.A., 61, 169, 254
Mt Monster, S.A.,
  welded joints, 83
Mt Painter, S.A., 6
Mt Whitney, California, 1, 17

Mt Wudinna, S.A., 145, 318, 319, 230, 321, 324, 325, 326
Moye, D.G., 156
'mud crack', 88
multicyclic landscape, 207 *et seq.*
    examples, 207-208
multiple flares, 244, 249, 252
Murphy's Haystacks, S.A., 236, 237, 273
mushroom rock, 247, 253, 254, 333
Myers, W.B., 29

Nanutarra, W.A., 12, 84, 166
Naraku, Q'd., 14, 266
Needles, The, S. Dakota, 177
nerviciones, 83
Nesbitt, H.W., 65
Neue Smitsdorp, R.S.A., 315
New England, N.S.W., 51
New plateau, W.A., 189
Nielson, R.A., 16, 115, 175
night well, 215
Niles, W.H., 156
nodular structure, 113
Noldart, A.J., 16
Nonn, H., 191
norite, 26
Norway, bornhardt, 137
nubbin, 14, 15, 16, 58, 124, 160, 161 *et seq.*
    evolution 164-167
nuclear explosion - and granite weathering, 61

Oberlander, T.M., 262, 299, 337
Obst, E., 125
offloading (pressure release) - problems, 152 *et seq.*
offloading joint (see sheeting joint)
old land, 207
Old plateau, W.A., 189
Olgas, N.T., 124
Ollier, C.D., 71, 111, 119, 124, 151, 155, 181, 280, 281, 301, 302, 309, 334
onion skin weathering, 111-112
Opferkessel, 215, 217, 226
Öpik, A.A., 207
order of weathering (minerals), 73
organic slime, 240, 251
oricangas, 217
Orme, A.R., 62
Ormerod, G.W., 113, 214, 215, 217, 218, 226
orogen, 27, 28
orthogonal joint set, 2, 7, 43, 44, 46, 51, 89
Osborn, F.F., 48
overhanging sidewalls, basins, 222, 225

overlapping slab, 321 *et seq.*

Paarlberg, R.S.A., 70, 85, 122
Paige, S., 200
Pain, C.F., 155, 181, 281
palaeoplain, 187
Pallister, J.W., 205
Palmer, S.A., 9, 70, 99, 304
Palmer, J., 16, 115, 175
Palo Ubin, Singapore, 102, 110, 271
pan, 214, 216, 217, 224
Pão do Assucar (Acucar), Brazil, 13, 74, 75
Papua New Guinea, 177, 181
Parslow, G.R., 29, 38
Parthenon, Greece, 294
parted block, 305 *et seq.*
partings - and weathering, 82-88
Passarge, S., 7, 125, 137, 262
Paulshoek, R.S.A., 165
Pearson Is., S.A., 6, 18, 52, 137, 248, 284
pedestal, 213, 228-232, 235
    definition, 228
    examples, 228
    origin, 228-232
pediment, 23, 190 *et seq.*
    apron, 191
    cone, 190, 191
    definition, 190
    'fan', 190, 191
    mantled, 14, 20, 21, 190 *et seq.*
    origins, 199 *et seq.*
    reasons for development on granite, 200
    rock, 193-199
pediplain, 22, 204
Peel, R.F., 200, 206, 207, 286
Peella Rock, S.A., 222, 258
pegmatite,sill, 80, 81
Penck, A., 281
Penck, W., 150
peneplain, 22, 188, 201 *et seq.*
    examples, 201-203
penitent rock, 96
Pennines, England, 257, 286
perched block, 95, 103, 104
perched boulder, 95
permeability, granite, 39, 40
Perry, R.S., 298
Pertnjara Hills, N.T., 177
perviousness, granite, 39, 40
Peterson, J.A., 318, 327
Pettijohn, F.J., 17
Peyro Clabado, France, 105, 288
phacolith, 31, 37
physical weathering, 60-64
pias, 217
Pic Parana, Brazil, 125, 130
piedmont angle, 149, 190-191, 265-267
piedmont nick, 190

pilancones, 217
Pilbara, W.A., 12, 16, 158, 160, 165
Pildappa Rock, S.A., 19, 78, 194,
   196, 197, 213, 215, 216, 220,
   242, 249
pillar, 98
Pirsson, L.V., 36, 93, 109, 318, 327
pit, 77, 214, 216, 223, 224
Pitcher, W.S., 29, 30
pitted granite, 62, 77 *et seq.*
pitting, 274
   distribution, 77
   in gneiss, 77
   durability, 78, 79
plain (granite), 5, 17, 22, 186 *et
   seq.*
planation surface, 207
plant roots and acid weathering, 66
plateau forms, on granite, 59
platform, 20, 21, 143, 145
   rock (pediment), 193-199
   structural, 28
Playfair, J., 36
Plenderleith, H.J., 294
plutons, 29
   classification, 29 *et seq.*
   S.W. England, 34, 45
pocos, 217
Poldinna Rock, S.A., 258
polygenetic forms, 334
polygonal cracking, 306 *et seq.*
   on platform, 308-310
   in subsurface, 315, 316
Pomona quarry, Zimbabwe, 143
Poondana (Brazil) Rock, S.A., 212
Popoff, B., 291, 296
pop-up, 318
porosity, granite, 39, 40
positive feedback, 94, 133
pot hole, 2, 214
Potts, A.S., 64
Pouyllau, D., 65
Powers, W.E., 2
pressure release, 152 *et seq.*
   and exfoliation, 112
   in relation to boulders, 112
pressure release joint (see also
   offloading, sheeting joint)
Price, N.J., 134, 328
Price, W.A., 316
products of weathering, 65, 68
protection, by duricrusts, 59
pseudobedding, 48
Pseudokarren, 237, 271
Pugh, J.C., 138, 199, 200, 262
Pumpelly, R., 108
Pyramid Hill, Vic., 27
pyramidel peak, 179
Pyrenees, 17, 63, 72, 76, 77, 83,
   177

quartz, weathering of, 67
Quincy, Mass., 51

Radley, J., 16
Rahn, P.H., 190
Ranc de Bombe, France, 171
rate of development of basins, 219, 220
Raupach, M., 66
Raynal, R., 191
Read, H.H., 27
regolith, 59
Reid, C., 109, 220, 279
reinforcement effect, 94, 133
relationship peneplain and pediment,
   202-204
relief inversion
   and doughnuts, 234, 235
   fluting or grooves, 240, 263, 271
relief of load joint, 49
Remarkable Rocks, S.A., 161, 162, 169,
   272, 284
remnant of circumdenudation, 138
Reusch, H.H., 291
Reynolds Range, N.T., 60, 107, 157, 169
ribbing (tafoni), 290
Richey, J.E. 51, 55
Richter, G.D., 45
rift, 46, 48, 49
rift valley, 1
Rille, 237
rims, 86, 87
   gutter, 237
ring complex, 33
Rio de Janeiro, Brazil, 7, 15, 106, 182
Rio harbour, Brazil (see also Bahia
   Guanabara), 7, 13
Ritchot, G., 167, 173
river film, 298
Roberts, N.K., 294
roche moutonnée, 2
rock basins, 17, 19, 77, 213, 214-228
   and Druids, 217
   and fractures, 215, 216
   and moisture attack, 219
   distribution, 214
   in other rock types, 214
   morphological differentiation,
      221-227
   nomenclature, 214-217
   origin, 217 *et seq.*
   overhanging sidewalls, 222, 225
   rate of development, 73, 219, 220
   subsurface initiation, 219, 221
   topographic distribution, 220, 222
rock bason, 214
rock composition
   and weathering, 73-79
   variations, 127
rock doughnut, 213, 233-236
   definition, 233
   examples, 233

origin, 233 *et seq.*
rock hole, 214
Rock of Ages Quarry, Vt., 57
rock pediment, 20, 21, 193-199
rock platform, 20, 21, 193-199, 258-260
rock texture - variations, 127
Rocklin Quarry, California, 20
Rocky Mts, Colorado, 63
Rognon, P., 169, 337
Romanes, J., 23
Rondeau, A., 113, 294, 297, 298
Rooiberg, R.S.A., 171
Rooifontein Valley, R.S.A., 208, 209
rounding and weathering (inselbergs),
   149 *et seq.*
   of blocks, 91-105
   of inselbergs, 149 *et seq.*
Rowan, I.S., 154
Rundkarren, 267
runnel, 213, 237-242
Rupununi, Guyana, 280
Russell, R.J., 2
ruware, 9
Ruxton, B.P., 117, 151, 262, 265, 267

St Levin's, Cornwall, 102
Salinaland, W.A., 258
salt crystallisation, 292-297
Salzsprengung, 295
Sandford, K.S., 207
Sapper, K., 262
saprolith, 59
Sardinia, 285
Saussure, H.B. de, 60
Sbar, M.L., 318
scalloped river bed, 2
scalloping (tafoni), 290
scarp, stable, 200
scarp foot depression, 199, 260-265
scarp foot weathering, 199, 286
   and stepped forms, 209-211
scarp retreat, 138 *et seq.*, 181, 183
Scheffer, F., 299
Scherber, R., 295
Scheidegger, A.E., 152, 157
Schmidt-Thomé, P., 227, 270
Schneider, E., 177
Scholtz, D.L., 271
Schrepfer, H., 173, 299
Schuchert, C., 93, 109, 318, 327
Schumm, S.A., 146
Scilly Is., England, 8, 74, 258
Scotland, northwestern, 187
Scott, W.B., 318, 327
Scott, G.D., 237, 269
Scott, G.R., 193
scratch circle, 298
Scrivenor, J.B., 92
sea fogs, and salt weathering, 295
Selby, M.J., 125, 138
Serra da Mantiqueira, Brazil, 17

Serra do Araras, Brazil, 17
Serra do Mar, Brazil, 177
Serra do Sol, Brazil, 178
Serra dos Araras, Brazil, 177
Seurin, M., 65
Shaler, N.S., 43, 49, 84, 108, 113, 150
Shannon, C.H.C., 280
Sharp, R.P., 191, 298
Shashe, Zimbabwe, 161
Shaw, P., 280
sheet structure, 2, 17, 18, 20, 47, 49
   et seq., 127
   age, 55
   and faulting, 52, 57, 155
   and glaciation, 152
   chemical weathering, 150-151
   disintegration, 119-121
   endogenetic origin, 154 *et seq.*
   exogenetic origin, 150 *et seq.*
   insolation, 150
   lateral compression, 155 *et seq.*
   metasomatic expansion, 154
   offloading, 151
   pressure release, 151
   plutonic injection, 154
   relation to other structures, 55
   steep dip, 52, 54
   subsurface disintegration, 120-122
   vertical uplift, 154
sheet tafoni, 286
sheeting joint (see sheet structure)
shell, 49
shelter, 286
Sherman Surface, U.S.A., 173
shield, 27, 28
shore platform
   Eyre Peninsula, S.A., 2, 5, 8, 25
   in granite, 2
sial, 26
sidewall tafoni, 286
Sierra Nevada, California, 1, 2, 5, 6,
   17, 177, 209, 210
Sierra Nevada batholith, 30, 32, 40
significance of time
   in granite morphological development,
   335
silcrete, 147
   on granite, 59
silica stalactite, 281
Silikatrille, 237
sill, 33
sima, 26
Simonen, A., 30, 38
Sinai Peninsula, 17, 177
Singapore, 23
Skertchley, S.B.J., 108
slab, 96
slickensides (on sheeting plane), 156
slipped slab, 318, 320, 321
slope behaviour, 138
slot (see cleft), 87

Smalley, I.J., 292, 294, 297
Smith, B.J., 286
Smith, D.L., 227
Smith, J.M.B., 318
Smith, L.L., 214, 226
Smithson, S.B., 29
Snowy Mts, N.S.W., 92, 112, 113, 115, 156, 179, 316
Soen Oen Ing, 137, 147, 177
soil moisture, 81
solubility of minerals, 65
solution, 64, 116
Sosman, R.B., 60
spall plate, 49
Sparrow, G.W.A., 16
specific gravity, granite, 39
spheroidal weathering, 93
split boulder, 301
    development, 304
split rock, 301
    development, 304
    and insolation, 302
spire, 95
Spitzkoppe, Namibia, 6
Sprigg, R.C., 10
Springbok Flats, R.S.A., 204
stability of scarps, 200
stalactite, silica, 281
Stapledon, D.H., 119, 152
star fracture, 309, 311
Steers, J.A., 207
stepped forms, 207 et seq.
stepped inselberg, 144 et seq.
stepped topography, 208, 210
stock, 31, 35
Stockbridge, H.E., 108
Stone Mt, Georgia, U.S.A., 125, 127, 130, 157
Streaky Bay, S.A., 8, 25, 114
Street, A., 67, 74
Streckeisen, A.L., 36
stress in anticline, 134
stress measurements, 156
stress release, 112, 113, 327-329
stretching plane, 49
Stewart Ridge laccolith, Utah, 35
structural control of granite relief, 24
structural convergence, 334
structure and castle koppies, 168
structure en gros bancs, 49
structure of earth, 26
structure, significance in granite morphology, 336-337
subsurface initiation, 24, 334
    boulders, 89 et seq.
    bornhardts, 135 et seq.
    and palaeoclimate, 136, 335
    minor forms, 143
    flutings, 275-279
    flared slopes, 144, 145, 245-247-257

    gutters, 275-279
    rock basins, 219-221
    tafoni, 145, 253, 288, 291
    polygonal cracking, 315, 316
    scarp foot depressions 263-265
    piedmont angle, 265-267
subterranean water, 24
Suess, E., 207
sugarloaf, 13, 15
Sugarloaf, The, Brazil, 13, 75
Sugden, D.E., 218
Sullivan Rock, W.A., 143
surface d'aplanissement, 207
Sutton, D.J., 45, 328
Sved, G., 317
Swakoprivier valley, Namibia, 141
Sweeting, M.M., 316
syenite, 39
Syers, J.K., 234, 271
Sykes, G.R., 318
synform - in sheet structure, 53, 55

Taber, S., 295
Table Mountain Sandstone, 187, 188
tafone (see tafoni), 281 et seq.
tafoni, 2, 17, 20, 23, 145, 273, 280, 281 et seq.
    and flared slopes, 291
    and joints, 285
    basal, 235
    boulder, 286
    initiation, 288-292
    sheet, 286
    sidewall, 286
Taj Mahal, India, 294
Talobré, J., 112, 156
talus cone, 72
Tamanrasset, Algeria, 194
Tampin, W. Malaysia, 10, 60, 62, 79, 268, 278
Tanner, V., 187
tanque, 215
Tarr, R.S., 48
Tarr, W.A., 60
Tassili Mts, Algeria, 100, 161, 336
Tatra Mts, Poland, 281
Tcharkuldu Hill, S.A., 308
Tedford, R.H., 184, 278
Tenaya Lake, California, 55, 153
Terzaghi, K., 328
tetrahedral cornerstone, 111, 114, 116
texture and weathering, 29-82
thermal expansion, 60
Thiele, E.L., 137
Thomas, M.F., 16, 23, 71, 94, 137, 152
Thompson, J., 294
Thompson River Valley, Colorado, 158, 159
Thorbecke, F., 262
Thorp, M.B., 125
Thury, H. de, 294

Tijuca Massif, Brazil, 7, 177
Toit, A.L. du, 33, 37, 120, 128, 146
Tolmer Rock, S.A., 231, 232
tombstone, 96, 107
Tomich, S.A., 139
tonalite, 39
Tooma Dam, N.S.W., 113, 115
tor, 16, 81
torr, 16
Torrens Gorge, S.A., 152
Torridonian, 75
   sandstone, 187
tortoiseshell rock, 286
Tosa Gargantillar, Pyrenees, 170
transition from fresh-weathered
   rock, 111
Trendall, A.F., 151
triangular wedge, 156
Tricart, J., 138, 183, 295
Tschang Hsi-Lin, 214, 268, 271
Tuan, Yi-Fu, 200
Tuddenham, W.G., 124, 281, 309
Turner, H.W., 109
Turner, F.J., 67, 296
turret, 15, 95, 100
turris, 16
turtleback, 8
Twidale, C.R., 2, 15, 16, 20, 23, 45,
   48, 50, 51, 55, 59, 60, 65, 77, 114,
   124, 125, 134, 135, 136, 138, 146,
   147, 149, 153, 154, 155, 172, 177,
   181, 183, 184, 189, 190, 191, 193,
   199, 200, 201, 205, 206, 214, 215,
   219, 226, 227, 228, 233, 234, 236,
   238, 245, 247, 258, 260, 262, 263,
   265, 269, 275, 276, 277, 278, 279,
   281, 286, 288, 291, 298, 303, 208,
   312, 317, 321, 322, 327, 328, 329,
   333
two stage development
   boulders, 94, 96 et seq.
   bornhardts, 135 et seq.
twr, 16
Tyrrell, G.W., 150

Ucontitchie Hill, S.A., 20, 21, 52, 54,
   56, 145, 194, 244, 245, 284, 272,
   274
   fracture density, 132, 133, 141
Ukraine, U.S.S.R., 27
Ule, W., 271
Urbani, F., 269
ultiplain, 22, 23, 204, 207
unconformity, 187, 188
undercut sidewall (of gutter), 237
Usakos, Namibia, 191
U-shaped groove, 267.

Valley of a Thousand Hills, Natal, 139,
   147
valley side facet, 208

Valverde, D., 131
variations in fracture density, 132, 135
Varley Township Hill, W.A., 259
varnish - organic, 299
   structure, 299
vasque rocheuse, 215
veins in grus, 93, 97, 98, 106
Velde, B., 64
Verhoogen, J., 67, 296
vertical wedge, 325-326
Verwitterungsnäpfe, 215
Veyrières, France, 246, 250
Vidal Romani, J.R., 83
visor, 286, 297 et seq.
Vogt, J.H.L., 52, 154
volume increase, and corestone
   development, 118, 119
Vredenburg, R.S.A., 271, 276, 277
Vredefort dome, R.S.A., 33
Vredefort quarry, R.S.A., 142

Wagner, P.A., 89
Wahrhaftig, C., 60, 151, 193, 208, 209
Wall, J.D.R., 268
Walther, J., 292, 295
Warth, H., 61, 314
Watanabe, T., 61
water as solvent, 64
water curtain effect, 273
water eye, 214
water scour - and doughnuts, 234
Waters, R.S., 48, 49, 81, 151, 207
Wattle Grove Rocks, S.A., 264
Watts, S.H., 73, 167, 218
Watts W.W., 37
Waulkinna Hill, S.A., 161
   pediment, 190, 194, 197, 199
Wave Rock, W.A., 245, 248
Wayland, E.J., 59, 187
weather pit, 214
weathering, 58 et seq., 186.
   acid, 65, 66
   alkaline, 65
   alveolar, 286-288
   biotite, 67, 74
   by moisture, 24
   changes in depth, 117
   chemical, 64-68
   defined, 58
   factors in, 71-88
   feldspar, 67, 68
   in geological past, 180
   initial, 68-71
   peripheral to corestone, 111-119
   physical, 60-64
   products, 65, 68
   progress on joint block, 117
   quartz, 66
   rounding of inselberg, 149 et seq.
weathering front, 59, 60, 189, 200, 335
wedge, 52, 56, 156

372

    lateral, 327
    vertical, 326, 327
wedge-shaped slabs, 52
Weed, W.H., 36
Wellman, H.W., 295
whaleback, 8
Whitaker, C.R., 20, 191, 302
White, R.E., 45, 328
White, S.E., 20, 71, 294
White, W.A., 149
White, W.B., 269
White, W.S., 152, 153
Whitney, J.D., 154
Wilford, G.E., 268
Wilhelmy, H., 23, 70, 135, 227, 276,
    292, 295, 297, 307, 333
Williams, G., 16
Williams, G.E., 75, 146, 187
Williams, I.S., 3
Willis, B., 8, 146
Wilson, A.T., 295
Wilson Promontory, Vic., 7
Wilson, R.C., 137
Winkler, E.M., 292, 294, 295
Witrivier Valley, R.S.A., 141, 163
Wojcik, Z., 281
Wolff, R.G., 317
Wolters, R., 152, 313
Woodard, G.D., 184
Wopfner, H., 147
Worden, J.M., 336
Worth, R.H., 48, 128, 214, 220, 227, 306
    306
Wray, D.A., 128
Wyatt, J.D., 16

Yarwondutta Rock, S.A., 145 (map),
    250, 252, 256, 264
    drainage system, 239
Yilgarn Block, W.A., 139
Yosemite region, California, 2, 6, 49,
    51, 129, 131, 137, 139, 153, 156

Zumberge, J.H., 295